Dieter Duhm

TOWARDS A NEW CULTURE

From Refusal to Re-Creation
Outline of an Ecological and Humane Alternative

About the Author:
Dr Dieter Duhm, PhD (sociology), historian, author and psychoanalyst, was born 1942 in Berlin, Germany, and is the initiator of the "Plan of the Healing Biotopes", a plan for global peace.

In 1967 he began his engagement in the Marxist left, and went on to become one of the leading figures in the students' movement of that time. In 1972 his well-known book Angst im Kapitalismus ("Fear in Capitalism") was published, which linked the thoughts behind political revolution to the thoughts behind the liberation of the individual.

In 1975 he started a three year social experiment with 40 participants in Germany. With the theme of "founding a community in our times" the experiment embraced the questions of the origin, meaning and aim of human existence on planet Earth. From this experiment, outlines of a new possibility of existence arose, with the concepts of "free love", "spiritual ecology", and "resonance technology".

In 1995, together with the theologian Sabine Lichtenfels and others, he founded the Tamera Peace Research Centre in Portugal, which today has more than 170 co-workers.

Dieter Duhm has dedicated his life to creating an effective forum for a global peace initiative that can match the destructive forces of capitalistic globalisation.

First edition, © 1993 Verlag Meiga GbR
Monika Berghoff • Saskia Breithardt
Waldsiedlung 15 • D-14806 Belzig
Tel. +49 (0) 33841 30538 • Fax: +49 (0) 33841 38550
info@verlag-meiga.org • www.verlag-meiga.org

Translated from the German by Claire-Lilith Suscens
and Sten Linnander

Original Title: Aufbruch zur neuen Kultur
ISBN 978-3-927266-37-7
Layout: Juliane Paul
Printed by Lightning Source Ltd. UK/USA

This book is freely available as pdf: **www.verlag-meiga.org**

Violence is the eruption of blocked life energies.
Pacifism does not mean the gentle appeasement of violence, nor does it mean brushing conflicts aside with appeals for peace. True pacifism is the radical and intelligent self-engagement of the human for the liberation of all life energies and creative forces that are present in him. Pacifism is the fundamental fight against every kind of suppression of human longing. Pacifism means taking an uncompromising stand for all living beings. Pacifism is militancy, not necessarily political militancy, but militancy in achieving inner truthfulness and freedom, for pacifism is the reconciliation of the human with himself.

Dieter Duhm

CONTENTS

Translator's Note 6
Publisher's Foreword 7

1 INTRODUCTION
Established Insanity 9
The Timeliness of the Concrete Utopia 10
Cultural Centres for New Basic Experiences 13
A New Consciousness of the Living 14
Overcoming Fear 17
Humaneness 18
The Bankruptcy and Rediscovery of Geist 21

2 THE EXISTENTIAL QUESTION
Behind the Ideological Façades 25
Wanderers in the Desert 26
A Different View of Suffering and Its Consequences 32

3 THE TRUTH OF THE LIVING
Fear as A Biological Disease of Culture 40
Life Research as the Science of the Future 41
Contact and Truth 43
The Technological Wonderland of the Living 44
The human in the Entire Organism of Nature 46
Functional Principles of the Living 48
A New Mental-Spiritual Attitude 53

4 BIOLOGICAL HUMANISM AS THE FRAMEWORK FOR A NEW CULTURE

The Concept of Biological Humanism	55
Three Steps towards a Realistic Humanism	58
An Ecological Relationship to all Living Beings	63
The Idea of Science	66
Evolution and Growing Freedom	68
A Culture without Sexual Repression	71
The Meaning of Sexuality	75
A New Social Organisation of Sexuality	80
The Question of Non-Violence	86
The Question of Democracy	90
Building a Humanely Functioning Community	95
Emotional Cleansing and Dissolving the Character Armour	98

5 POSTSCRIPT

Concerning Tradition	102
That All This Does Not Remain Mere Words …	107

6 APPENDIX

Tamera Manifesto	111
Thirty Years Later	123
Further information	125
Literature	125

TRANSLATOR'S NOTE

Some minor editing has been made to adapt the book for an audience unfamiliar with specific political and cultural developments in Germany in the 1960s and 70s.

For the sake of readability, the German word *Mensch* has generally been translated either as "man" or "human" (referred to as "he" and "him", rather than "he/she" and "him/her"), to be understood as including both genders, except where the context makes clear that specifically the male part of humanity is being referred to.

The author uses the term *das Lebendige* ("the living"), referring collectively to all that lives. This has generally been translated with capitalisation as "the Living". In a few cases, for clarity and readability, it has been translated as "the living realm".

The German noun *Geist* and its adjective *geistig* may be roughly translated, with some loss of subtlety of meaning, as "mind-spirit" and "mental-spiritual". In the translation that follows, *Geist* has been translated in some cases as both "mind" AND "spirit", and in other cases as either "mind" OR "spirit", with the choice of translation being in accordance with the context. For the most part, the adjective *geistig* has been translated as "mental-spiritual", although in some cases the words "intellectual", "philosophical" or "religious" have been chosen as more appropriate for the context. The exception, where the words *Geist* and *geistig* have been left untranslated, is in the title and introductory sentence of "The Bankruptcy and Rediscovery of *Geist*", as here it was important to retain the subtlety of meaning of the concept of *Geist* that is central to this section of the book.

PUBLISHER'S FOREWORD

How will life go on after the collapse of the globalised political and economic systems? How will it survive the large-scale disruption of our planet's ecological and climatic systems? And what will become of the immense systems of belief, love and thinking as they start to shake within?

The answer to these questions has to persist in the face of so many failed attempts in the past. The world stands on the brink of an abyss.

The youth from Cairo to London, from Greece to Chile, from Rothschild Avenue, Tel Aviv to Wall Street, New York are looking for new ways out of the crisis. If the mass protest and revolt movements rising up all over the world today intend to combine their revolutionary power and take off together, if life is to win over violence and war, we need a direction, an image, an idea of what might be our common goal.

This book offers an idea of how a future worth living could be. It was written and published more than thirty years ago in Germany and we believe that its time has now come. We left it in the political context in which it was written. Through this we want to show that names change, yet the underlying problems remain the same … until we discover how to solve them. How this can be achieved is what this book is all about. It is more relevant today than ever.

The author, Dieter Duhm, has given a voice to life itself here. He has tracked it behind false morals and dogma, and has opened up ways for it through the walls and armour surrounding heart and mind that we all needed in order to survive an epoch hostile to life under patriarchal rule.

But all this could be over now.

The system change that is taking place today, is the most profound and fundamental that has happened in thousands of years. It is a change from the power to destroy life to the power to care for and protect life. This is the only way for this planet Earth and all its inhabitants, including the human being, to have a chance for a future.

We wish for this book to meet open ears and hearts, and that its seeds of humaneness and compassion will bear fruits worldwide.

This is more than a book. It is an idea of how a future worth living could be. The author has taken himself at his word and set out together with comrades and friends to put this idea for the future into practice. The new edition – published for the first time in English – includes an appendix section that shows in brief, what now has resulted after thirty years of pioneering work. A dream is becoming reality …

May this undertaking succeed, because "if life wins, there will be no losers".

Monika Berghoff
Publisher
December 2011

1 INTRODUCTION

ESTABLISHED INSANITY

The political and civil insanity of our epoch has reached its climax. An inner, systemic, human process of destruction has robbed the heart, sense and reason of modern societies. Armed to the teeth, and directed by incorrectly programmed hierarchies of power and money, they launched an attack upon life on Earth that can no longer be stopped by conventional means. The evolution of humankind has reached such a fantastical impasse that its dimensions compare with those of a utopian novel. We stand at the end of an era and perhaps at the beginning of a new one.

The compelling political, economic, ecological, social, technological, psychological, medical, scientific, and mental-spiritual questions of our time need different answers than those available within the framework of our established ways of living and thinking. The analysis of all of these questions leads to the same realisation: they can only be solved truly and with lasting effect if the human finds a fundamentally new relationship to himself, to his fellow humans, to all other living beings, and the entire planet. A "new relationship" means a new behaviour, a new way of life. In the heart of a new approach towards an ecological culture stands a liberated, unsentimental, and active love relationship with all that lives. It creates the opening that makes everything else visible and understandable. The interconnected processes of alienation, mechanisms of destruction, and disorientation of our time has become so total that, in a sense, we must start all over again if a perspective for the humane survival of the human species is to arise. The political and ideological systems that guide today's societies no longer have the qualifications necessary to prevent disaster; they have become fatally incompetent. From now on we must step into the position where we are able to take responsibility ourselves. We may then have the hope that our energies will exponentiate if we use them wisely. In terms of new ecological politics it makes more sense to count on the effect of interpersonal resonance than on a change of power in the State or the economy.

All the admonishers of our time agree that we need new human values. Often enough, though, we have experienced that appeals to reason and conscience have no effect. Even the best of words are already so ruined that we can hardly use them any more. New humane values will only gain substance and reality when they are anchored in a new way of life that opens up new energies, contacts, fulfilment, and realisations for humankind. This search for such new life structures is the true adventure of our times. Since this search has been occurring dispersed throughout the nooks and crannies of our society its historical importance is easily overlooked. The New arrives on quiet feet. But it has, in the meantime, amassed so much experience and knowledge that we can decide to make a more comprehensive start.

THE TIMELINESS OF THE CONCRETE UTOPIA

The killing every year of over 200,000 seals for the production of fur coats… the poisoning of living waters that are the source for all that lives… the daily psychic murdering of children and youth in a system of dishonest and anonymous relationships… the suffocating of the Living behind concrete walls of bureaucratic indifference and pretence… the annihilation of farmers and freedom-fighters in the Third World…

The precision and sterility of our civilisation is connected to such an extent with the agony of murdered life, that daily more and more people resolve themselves to step out of this type of morality and sterility and to stand against the crimes of the established system. The groups and initiatives that grew from this worldwide defensive struggle – from citizens' initiatives to occupations of nuclear power plants – from Greenpeace to Amnesty International – would be enough in number to change significantly the entire situation on Earth if they had succeeded in directing their resistance into stable and constructive paths (as happened briefly and analogously in the establishment of villages and communication networks on occupied sites).

Out of the structures of defence and resistance, we must build and develop new, positive cultural and societal forms. Otherwise, the success achieved in life attitude, solidarity, and readiness to act will fall victim to the *inner human mechanisms of destruction* of the existing systems, as happened, for example, after the students' movement of the 1960s and 70s in Germany. One thing is clear: it was not State repression that brought the German "New Left" to its knees after such a short time, but the ideological and psychological destruction stemming from the unsurmounted inner emotional structures of conflict. The success of our efforts to defend life and the Earth will stand or fall in the long term on our being able to develop our *own* new, concrete cultural and social approaches for our *own* future *in which the psychic mechanisms of destruction between people can be fully recognised and brought into a process of resolution.* A new cultural approach that does not take this problem into consideration is in fact no approach at all. Our experience tells us that it will fundamentally maintain the old structures regardless of how revolutionary, spiritual or ecologically interesting the terminology may sound.

Here we have reached the most critical point of the current alternative movement. A new way of life means new ways of working and living, new kinds of nutrition, and above all new forms of love, of community building, of sexuality, of resolving conflict, and a new social transparency, especially in all areas that have so far remained private. More group projects have failed because of unseen and unsolved events in the areas sexuality and jealousy than because of economic or political difficulties. A new culture will be rooted in, among other things, a new relationship between the genders. Eros is surrounded by distrust, hate, and jealousy; sexuality is caught in a web of projections, fears, and pretence. Hardly a direct word, a clear gesture, or an open contact is possible. There is a common resentment – which has spread especially in the political left and counter-cultures – that has turned people away from the impulses of the student movement that leaned more towards community and free sexuality. The problem of building humane, functioning communities, free of unsolvable conflicts of jealousy and power, is associated with so many bad experiences that in the related themes involved in building new life structures, attention to these areas is almost entirely missing. However, these problems constitute the decisive factor. Whether or not we have a future worth living hangs largely on their resolution.

Perhaps the old concept of the "Concrete Utopia" has never had such an historical, political, and psychological timeliness as it does today. The draft of a desirable future, one that a growing number of people will find plausible and realistic, is the number one priority on today's agenda. Tens of thousands of people would be ready to take part in such experiments the moment its positive possibilities become known. To "drop out" would then mean to "drop into" something else, to choose something more meaningful and more rewarding. But such a breakthrough towards a concrete utopia – such a joyful unity of willpower towards a total cultural reconstruction – cannot succeed as long as the unsolved "human problem" remains systemically in the way.

CULTURAL CENTRES FOR NEW BASIC EXPERIENCES

Because humans are human, they will, in the long run, only embrace a new culture if it meets, acknowledges and accepts the "neediness" of their bodies and souls. Only then will they be able to devote themselves fully to things that transcend their direct needs, without a false aura of holiness. By disregarding this basic principle, practically every attempt at humanisation so far has failed.

Such simple truths, once spoken, usually generate a turmoil of questions and doubts. Suddenly the true fears and problems become visible. What will happen to the obese, the unsightly, the disabled? Who accepts the compulsive neurotics, the chronic complainers, and what kind of affirmation could sexual perversion receive?

To be viable and "in tune" a new culture will need a clear living concept that provides satisfactory answers to these and many other such questions. Developing such a concept is not a desk job, nor will it come from the experiences of traditional society. It cannot use the old words, nor will it add any new ones to existing theories. It can also never be finished. It is already emerging in places where people are having substantially new experiences, with themselves and with others. It is emerging at several points on Earth, though still in a fragmentary and contradictory way – in places such as Findhorn in Scotland, in some centres of the Sannyasin movement, or in Friedrichshof* in Austria (a therapeutic and cultural centre that emerged out of the AAO movement). No matter how bizarre and

---

* Editor's note, please be aware: The statements on all projects are related to the time thrity ago (1979), when this book was written, and do not contain any judgement about their later development.

shocking these experiments may seem to the rest of society, no matter how little people may agree with them, they are true workshops of a new human culture. Any organism undergoing true birth pangs tends to act in strange ways. These places must be experienced first-hand to see what they are essentially about. Without doubt their essence is at a level beyond verbal debate; it is about new experiences where popular questions of power and authority, of individuality and autonomy, of democracy and non-violence, and above all of morality and sexuality, have to be separated radically from all the old intellectual games of habit. They must be viewed in an entirely new light based both on one's own experience and the communal experience. A new cultural concept arises from a new way of seeing things. This new way of seeing is the result of new basic experiences. *The urgent task for the alternative movement is to establish inner focal pillars and centres where such basic experiences for a new culture can take place.*

A NEW CONSCIOUSNESS OF THE LIVING

From the confusing fragments of our age different lines are forming a new synthesis. They converge towards a new consciousness of life. Life, both external to and within humans, of which all soul processes are part, represents the two sides of a comprehensive, universal, sensory and extrasensory world comprising all that lives. This world of the Living, in which everything human and social is embedded, is like an integral organ of the total organism that is either functional or dysfunctional. The special structures and functioning principles of this organism contain a multitude of messages about other possibilities for human life (see Chapter 3), about other methods and goals for self-realisation, creation of social structures, for medicine, architecture, and technology in general.

At the heart of the present cultural crisis is the collision between the working principles of modern society and those of the living realm, or, in short, the collision between sociosphere and biosphere. This collision takes place both outside and inside the human organism. It can be seen in the dead lakes and rivers as well as in the epidemic of identity disorders and psychosomatic illnesses. Ecological and emotional misery are two aspects of the same cultural dead end street. A convincing cultural concept must, today, be able to address both areas. The main theme of the coming era will be that of integrating human life into the universal world of the Living. This needs to take place on all levels of our existence, from the sexual to the technical, and the political to the spiritual. It is the common theme for all cultural centres that are currently develop contributions to the society of the future.

As things stand today, ecological restoration work and attempts to protect and benefit the Earth for the long term means, of necessity, working outright on the human being. This work needs to be seen in a much deeper context than is suggested by the slogans of "anti-consumerism" or "environmental awareness". The crucial events for a new future are taking place in the inner zone of our lives: in the erotic area, in the areas of cognition and religion, and in the form of our daily lives together. The deeper we get into these areas, the more they overlap around a central core that shines through them all. To rediscover this core at a new level of consciousness and to cultivate the Earth from this basis may be the quietest, yet greatest theme of the new developments. All great processes of change are spiritual in nature, even though we are forced to address them in their crudest and most laborious aspects.

If the human is reconnected to his centre and therefore to universal life functions and consciousness processes, then a new type of natural life-affirming humanism will emerge of itself. Human properties that emerge of themselves through such inner integration are, for example: strength without domination, firmness without hardness, clarity without coldness, tension without stress, posture without rigidity, softness without weakness, beauty without vanity, adaptability without self-sacrifice, and charisma without manipulation.

The world of the Living is full of contradiction. The "functional unity of the contradictory" (see Chapter 3) is one of its operating principles. This living contradictoriness also runs through human consciousness where it is expressed in opposing theses, world-views, liberation theories, and so on. A life-oriented cultural concept will not take part in the old battles between opposing positions. Instead it will, without compromise, assimilate the truths of each belief. The resulting mental tension and suspense must be endured. Finding premature "solutions" to spiritual or mental problems by reducing the total truth to one single aspect was an ideological method of the old era. In the history of wars based on ideology and beliefs, that attitude led to so much bloodshed that it must be exchanged for a method more in tune with life. One of the greatest polarisations in the alternative movement today is that between the emotional and spiritual concepts of life renewal. Let us take, for example, Wilhelm Reich and Rudolf Steiner. The contradiction runs as deep as the truths inherent in each pole. Wilhelm Reich would liberate and heal the human by freeing creative energies in the emotional and sexual areas. Rudolf Steiner proposed to liberate and heal humans by freeing their creative powers in the mental-spiritual realms of cognition, thought, and ideas*. These are two cross-sections of truth about humans, two standpoints, two world-views. One is as well formulated and as irrefutable as the other. They no doubt belong to the most important ideas of our times, and yet they are alien to each other, almost hostile. In the scope of their differences and in the span of their varying approaches lies the depth of their possible convergence. This is the task of the present time. The synthesis that is called for here is that between body and spirit, between sensuality and religion, and rarely has such a synthesis so far been successful.

OVERCOMING FEAR

Such a synthesis of consciousness does not evolve out of a theoretical construct. Rather it comes about through the experience gained when overcoming engrained limitations which naturally require the readiness to expose oneself to such experiences. Ideological barriers are usually barriers to experience and at their psychological core they are almost always barriers of fear. Facing possible cultural doom, a cultural synthesis of the kind necessary today first requires the creation of a living environment where people can learn to live together without fear and thus without the usual barriers to experience. Do we know what that means? It means being able to show sympathies without fear of hurting anyone and without fearing revenge from anyone, being able to discuss with one another without fearing that anyone is competitively collecting adversarial points. Creating a psychological and social environment for overcoming fear is in itself a cultural task that violates all rules of proper conduct, good manners and customary habits. Fear binds together and ferments the customary forms of culture and society. Every serious attempt to overcome fear, therefore, violates the rules and taboos that are accepted and

---

* The anthroposophical concepts of thought and of the idea constitute an important part of a tradition in the history of ideas that until today has received little attention and been poorly understood. Plato's "Idea" or Goethe's "observing power of judgement" (anschauende Urteilskraft) are also a part of this tradition. Here realisations and discoveries have been made concerning the nature of perception, realisations that are of existential importance and in that sense follow the ideas of Teilhard de Chardin, who saw the development of an increasing cognitive ability at the heart of evolution.

observed as faithfully by bourgeois society as by the most dissident subcultures. Until we overcome fear, however, we can have no humaneness, no integration into the Living, no realistic solutions to any of the current problems that face us. Fear is life that has been blocked off; it closes off organisms, obstructs contact, and thereby lowers cognitive ability.

Overcoming fear, falsehood, and hiding from one another would mean, among other things, the liberation of sexual energies from all the cages in couple relationships, both in marriage and morals, and the complete liberation of all the repressed libidinous energies and life potential and their conscious integration into public social life. This sentence expresses one of the key points of this book and will reappear in various contexts.

HUMANENESS

Humaneness – what can it possibly mean today, considering all the atrocities that have occurred, in the face of the worldwide madness in which we are living and in front of the troubled future that, in spite of everything, we still want to win for ourselves? Humaneness – in order for this word to ever regain meaning and value, it must be freed from all the romanticism and sentimentality it has taken on during the era of the Church, of repressive morality, of the "beautiful soul" and of general hypocrisy. Humaneness became the badge of peace, hung around the neck of the human monster to protect it and its surroundings from its monstrous impulses. The label was so superficial that it could hardly unite with the flesh and blood of the brute underneath. If we still need proof of this, it has been given to us and is still being given a thousand times, in fascism, in the Vietnam war, in the murder of Brazilian Indians, and in the inhumanities reported daily in the newspapers.

Humaneness means *knowledge* of the entire human phenomenon, including its abhorrent dimensions, *accepting* the entire human phenomenon, and *changing* the entire human phenomenon. Knowledge! With dizzying consistency history has revealed the connection between ignorance and barbarism. Enlightenment, science, and the expansion of consciousness are – at least in concept – basic elements of any true humanism. In light of all lost illusions of freedom and brotherhood, democracy and socialism, reason and responsibility, the work of enlightenment must now concentrate on the human being, so that the creators of the new culture know what they are dealing with. In later chapters we will see the extent of all that must be acknowledged.

And furthermore, what does it mean to accept the human phenomenon, or to even have the ability to accept it? To accept does not mean to say yes with weak knees. To accept means to see and yet stay calm without succumbing to fear or hatred. There is so much terror, so much personal fear, so much distrustful bone marrow to be brought to light, lived through, assimilated, and worked out! Are not we, who have begun to regenerate ourselves, still too weakened by our own fears and lamentations to truly accept humans (for example, one's so-called love partner) and to say yes to whatever is needed? Here, as in so many areas of human affairs, Nietzsche looked deeper than anyone else.

And finally, what is meant by changing the human phenomenon in its entire scope? It is perhaps indicative that the desire for change so typically encountered in the Marxist left, in spiritual circles or with the strict raw-food mentality, when seen from a psychological standpoint, often turns out to be an ideological bastion against true change. So far, no religion and no political theory has suspected the extent, scope, and quality of the changes necessary for meaningful survival. It may be that every liberation movement so far has, at heart, been a campaign to divert attention from those secret inner places where the true longings, the greatest fears, and the most gnawing questions lie hidden. Politics, morals, and religion can be seen as distracting manoeuvres. We must also recognise this aspect in them, although of course it is not the only one.

Humaneness is the transformation of the animalistic conditions of existence into humane ones, *without separating the humane from the animalistic and pitting them against each other*. The totality of evolution, at work in humans in the form of their inner driving forces and paths of development, can be restructured and refined through new experiences and the power of thought without simply excluding large parts of life (for example sexuality and aggression) – this is an aspect of humaneness of interest to us. This type of sublimation process – if it takes place consciously and without any neurotic manoeuvres of repression – constitutes the essential process of becoming human. But in order to enable sublimation without hypocrisy or self-repression, how much raw material must first be dug from the murky depths to be transformed by the light of day? If today we wish to become humane we cannot simply start off humanely, for too much inhumanity, too many contortions, and too much dishonesty have become second nature to us. Our goal should never again simply be superimposed upon reality. Reality, in its more subconscious psychological zones, should first be able to expose and express itself in its current state of being. If this alone were achieved much would change. As this is impossible in our conventional system of communication in daily life, we need *experimental cultural workshops* for such learning processes.

Humaneness does not lie outside our instincts, drives, longings, or hunger for life. It lies within them. It does not lie in the aseptic isolation of the mystical East nor in the superimposed energy of universal love – as long as the underlying emotional structures are the same. Rather it lies in the driving energies of our body and soul and our love, once they have been grasped, brought to light, integrated, reconciled and refined. Humaneness does not lie on some island of beauty but in working on our culture and society, in working on oneself, in creating the conditions for one's life and in taking on political and overall human responsibility. Lastly, humaneness lies in discovering the common and universal aspects of oneself and consequently beginning to work with ever widening horizons.

The humane enterprise, not blurred by any pious slogans or fashionable fads, will realise some central ideas without which a future humanism probably cannot be conceived. For example, the idea of a non-violent society, the idea of grass-roots democracy, the idea of a culture without sexual repression, the idea of science, the idea of evolution and the freedom of humans in evolution, and the idea of community in solidarity with all living beings. We will see that these ideas, sounding as much like old hat as does everything that is worth talking about, will acquire a new meaning in a new context. Humaneness is the unredeemed part of worn out words.

THE BANKRUPTCY AND REDISCOVERY OF GEIST

It is difficult to speak of the "*geistig*"* at a time when even the most determined have discovered the false glibness that almost always lies behind such talk. We have peeked behind the façades of the representatives of mind and spirit and have become sceptical. There has rarely been a plea for the mental-spiritual nature of man that did not, consciously or not, have as its aim the distraction of people from their untamed animality and the human problem with humaneness.

---

* The German noun Geist and its adjective geistig may be roughly translated as "mind-spirit" and "mental-spiritual". See the Translator's Note for further clarification.

The philosophical, spiritual, and especially occult worlds are surrounded by the smell of putrid emotions and neediness, seething in dark and stale dungeons. Truly mental-spiritual forms of communication are rare in our time, since the beautiful words, even before they have been uttered, are already at the service of something that we may refer to as "lies of life". They are therefore not believed. A general lack of credibility today causes the power of thought, speech, theory, and spiritual inspiration to vanish from public exchanges between people.

Nevertheless, the budding new culture will, as an answer to the materialistic era, increasingly focus on mental-spiritual processes. Certainly the emerging cultural revolution will be a biological one, because it concerns the regaining of our outer and first of all inner sources of life and is about shaping our environment in accordance with the laws of organic processes. This book serves the single purpose of giving content to that statement. But that statement will be meaningful and supportive of our historical development only to the extent that the other side is also considered: namely that the coming cultural renewal must also be a mental-spiritual one. Neither statement has priority, they are complementary and belong together. Without changing and raising the mental-spiritual level, the biological sources of life can neither be found nor made accessible. But by finding and gaining access to them we shall discover with increasing clarity what many of us today are pregnant with: the mental-spiritual nature of life processes and universal energies. The discovery of the Living contains a spiritual experience, which in itself could build bridges between the different factions of our current world-views.

The power of resistance against pain and the capacity for its transformation into willpower are in part dependent on the power and certainty of a mental-spiritual orientation. The loss of mental-spiritual identity and orientation among many of our contemporaries leads to a passive condition that sooner or later results in overstepping the limits of resilience. This first of all is the source of the ravenous hunger for ideologies of salvation, a hunger that wishes to bypass development because there is no energy left for waiting and no feeling for the natural pace of development. An unrestrained lack of patience wants every thought and every spiritual insight immediately translated into "concrete" or "practical" reality. How fatal the terms "concrete" and "practical" are! How handy and how easily digestible a thought must be in order even to be heard! The often discussed relationship between theory and practice leads to a shallowness of the theory and a corresponding shallowness of the practice.

New fields, which could have heralded great things, are thus wasted in a peculiar manner, such as the fields of psychoanalysis, bioenergetics, ecology, and the new religiosity. They have hardly seen the light of day. Their depth and scope are not even visible when they are immediately used to enlarge someone's own little house even though they are actually designed for the scope of a world. Today we must rebuild the Earth.

Profound developments come quietly and at a slow pace. It is a good idea to develop a calmness in the midst of activism, especially as the earth can be likened to a barrel of dynamite. A calmness and meditation of seeing and perceiving, without immediately firing our whole barrage of goals, fears, desires, judgements, and accusations at the world. May it not be true that, after all the hullabaloo about our lost spontaneity has died down, the deepest energies, perceptions, and creations arise from a level of silence? We touch a thought, a chord, a love within us until it seems to touch the world. At the point of contact, there is the unmistakable experience of power, identity, and meaning. From these regions within us the ciphers of this world become brighter and more transparent, and all senses are sharpened by a sense of foreboding, expectancy, amazement, and thankfulness. Nietzsche and Teilhard de Chardin both attested to this.

2 THE EXISTENTIAL QUESTION

BEHIND THE IDEOLOGICAL FAÇADES

There is so much sadism, so much fear of women, so many unlived longings in the history of men, from the destruction of Troy to the neutron bomb. Why have we been able to cultivate war as a reality, but love only as a dream? There is so much hunger for a life never lived it has to be anaesthetised through alcohol, medication, or consumerism.

The inability to lead a fulfilled life leads to a loss of self-esteem. Through the loss of self-esteem we have lost a natural social integrity. humans today are adrift in the jungle of their own contrived forms of disguise and compensation. If more men were soulfully potent there would also be more women who were self-aware. The opposite is also true. Both men and women would like to love and surrender themselves completely, but to whom and to what?

The emotional structure of male culture today is a culture of seventeen-year-olds. Since the possibility of accepting woman as a partner has not yet been discovered, she is not taken seriously. Creating a fundamentally new intellectual, emotional, sexual, and social relationship between the sexes is the first necessary condition for the man to grow up and for the woman to be able to overcome her dependency.

The man's psychological power over the woman is based on his habit of indoctrinating her, ideologically and morally. The woman's psychological power over the man is based on his sexual fears and projections. As the patriarchal era had a habit of exorcising fears through indoctrination, the man won a shady victory in this power struggle.

A form of hypnosis and disorientation still dominates sexuality today. Sexuality devoid of fear and, therefore free sexuality, equals love; it does not have to be made into love through morality, special gentleness, or personal guarantees. If sexuality is not experienced as love, then the cause does not lie in the essence of sexuality, but in the blockages and perversions that stem from repression.

If you have been straightforward with yourself and one day realise that you feel no fear or guilt, have nothing to hide, and do not need to lie, then you are free. The liberation from deeply ingrained feelings of fear and guilt is the first prerequisite for a creative, energetic life, for being at one with life. The principle of fear is the most destructive and paralysing antithesis to the principle of love. But fear can be overcome by love. Then the individual will be prepared to go all out in life when he has found the path of his innermost desires, the path of which an Eastern sage said: "Tao is the way that cannot be abandoned; the way that can be abandoned is not Tao." Finding and pursuing that path causes change to occur, a change that can lead one radically away from all old habits and fears, and can also change the relationship to one's own suffering. Every sprouting bean destroys its own house as it seeks the light. In the power of his growth and his "becoming", the convalescent does not care about the destiny of his own skin. It is from this depth that the culture and politics of today need to be drafted.

WANDERERS IN THE DESERT

It took three billion years for life to evolve from the cell to the human being. That process sheds some light on the magnitude of what we are dealing with when we try to see the human phenomenon in a natural framework. Where do we stand today in the mysterious arc of development in becoming human, and which act lies ahead in this drama of creation?

Through our historical past we see a wave coming towards us that is gaining momentum at a terrible rate. Created some time around the seventeenth century, it is the wave of big industry, of population explosion, and of large technological wars. One could perhaps say that it is the wave of the globalisation of humankind in its first and most unconscious phase. Today this wave threatens to break. It has not yet broken, but may do so in the near future. Modern civilisation is reaching its boiling point. Further heating of its typical parameters of industrial growth, armament, environmental pollution, and emotional misery will result in a qualitative leap such as that which occurs when water transforms into steam. The evolution of humankind seems to stand before an imminent mutational leap.

History, in all its variations, has shown that when humankind embarks on a certain course, it goes all the way. This time going all the way would mean its own annihilation.

For the part of humanity wanting to survive in a humane and intelligent way, there is now one theme and one task: to co-create the transition from the cultural era based on the laws of profit to an era based on the laws of life. The two are usually diametrically opposed to each other. Therefore most of the institutions, organisational forms and habits of behaviour of the old era cannot be incorporated into the new one. On the contrary, overcoming these old ways is the pre-condition for a humane survival to be achieved at all. This need for change applies to the existing forms of economy, city planning, energy supply and landscape design. Furthermore, it applies to the whole area of technology, which with our already diminishing resources, will have to reorient itself towards entirely new materials and processes during the next 50 to 100 years. As a third and more important point, it also applies to the inner sphere of our lives, our emotional, sexual, and spiritual ways of living and our relations to each other. These are the areas where the human is directly connected to life from within.

How should this transition occur, and who will bring it about? Today, these questions have become so unanswerable that one could become cynical. What shall we do with the hypertrophied cities, the industrial monsters, and the traffic complexes if they are no longer needed in the coming era? Who – and with what political mandate – should build a decentralised ecological society at a time when the apparatus of production and distribution is more centralised than ever? Who should take power away from the mighty, from money, from corrupt "necessities"? Who will eliminate the power of social status, who will release us from the grip of our daily habits – including those of the so called opponents of the system?

Who? This question concerning the "revolutionary subject" can no longer be answered with sociological or political-economic theorems, because what needs to be revolutionised today is exactly that "inner part" of the human being and of society that can not be described in such terms. The necessary revocation of powers needed today is taking place in other arenas, not at those barricades where the political police and old-style revolutionaries bumble around. It is about the intellectual, spiritual, emotional, and libidinous disempowerment of bankrupt ways of life. Anyone who can substitute something better to counter the system is contributing to this disempowerment. For these activities not to remain in the private sphere, for them to unite meaningfully with other such activities and ultimately create a formative force for the future, a generally valid cultural and political concept that includes an overall alternative capable of development is needed. All those who are actively giving thought to our present condition are beginning to agree that in reality nothing is more important than the development of a positive, encompassing alternative, connected with a comprehensive and convincing idea for a new culture.

True enough... but... Now as we approach the heart of the matter, we see that such positive comprehensive concepts are practically non-existent. The Marxists live on nostalgia; the tide of therapy has stalled in individual problem-solving; and spiritual innovators are often so bizarre that their behaviour suggests mental-spiritual confusion rather than salvation. The environmentalists and activists from the counter-culture, as a rule, know much more about what they fight against than what they fight for. Here we should not fool ourselves. Non-violence, grass-roots democracy, and decentralisation are slogans and not yet manifestations of will. What does someone who wants grass-roots democracy really stand for – what does he really want – as a human? In a later discussion of grass-roots democracy (in Chapter 4) we will see how necessary it is to pose this question and how fatal to ignore it. *Only when people know themselves so well that they can say with certainty what they as humans want, and when they have liberated themselves from their ideologies to the point where they can freely communicate this, only then can they formulate cultural and political concepts that do not side-step their own true motives.* As long as they cannot do this, they are like two wanderers dragging themselves through the burning heat of the desert discussing political (or moral, or religious) questions. Suddenly they come upon water, and run off and drink like animals. *That* was what they had been missing. This happened to some powerful comrades, who served faithfully in the Marxist left until they had the courage to go to Poona (an Osho community in India) or to Friedrichshof (a free-love community in Austria). The thirst for water is the thirst for basic life fulfilment, contact, Eros, adventure, breathing freely, a meaning and purpose, creative realisation and self-esteem and for a life one can wholly affirm. Where are the cultural visions and the political goals for this kind of thirst? *Only for them would it be worthwhile to make a wholehearted commitment.* Transforming the elementary soul needs of the individual to the more general level of a new social and cultural concept is the most urgent task of our time.

Human and political comprehension is made difficult today mainly by the fact that we live in a hypnotic state. One example of this is our habit of giving false terms and explanations to matters that concern us in order to make them more socially acceptable. Added to that is our habit of believing those terms and explanations. One claims to move from the city to the countryside to escape the alienation and anonymity of the city. The truth is usually that our fears make us incapable of handling creatively the provocations and temptations of the large city. One pretends the need to be emancipated from social authorities and restrictions but there is only one true emancipation, and that is emancipating oneself from one's own insanity, projections and fears. One speaks seriously with indignation about the destruction of nature by humans, but suffers severely and lastingly from quite other things: problems of contact, an uncreative life, sexual difficulties, the inability to give oneself completely, depression, feelings of fear and meaninglessness, a lack of zest for life, and a constantly growing circle of psychosomatic illnesses with no specific causes. When developing a relevant cultural idea we must not let ourselves be deluded by the use of language designed to avoid real issues. We must be concerned, foremost, with providing an answer to the existential problems of present-day humans. *What is human, whatever that means concretely, is the true political issue of today.*

So we are concerned with existential matters. That is bad enough, because it has been about that before, and attempts to deal with existential matters have so often failed. There has probably been no time in history that needed this existential approach more than our time, and probably no generation that has rejected it as totally as ours has. When it comes to existential matters we have been burnt. The last major attempt, after Christianity, to change the world in an existential way took place under Fascism. The result was so devastating that since then, among more sensitive people, every word reminding us of truly existential issues has been taboo. The international students' movement, which, of course, arose from an existential impulse, availed itself of the pseudo-arguments of a political-economic theory. These were pseudo-arguments since none of the millions involved actually took to the streets for political-economic reasons. In reality what the participants wanted was a sense of community, a feeling of being alive, of solidarity, the courage to resist, and the feeling of personal competence. What happened at that time could be described much better in a psychological-philosophical language (if there were such a language, and if its words still had any meaning).

So we are dealing with existential matters – that means basic changes in the emotional, sexual, and mental-spiritual areas. In order to achieve that change we need a new set of priorities for how we view personal needs, a new life practice and a new organisation of daily life. Who, among those over thirty, would be prepared to implement that change? Verbal criticism of the system commonplace in the intellectual middle ground of society usually goes hand in hand with a conventional lifestyle. Why is this? Because the critics of the system are tightly bound. They have settled for a certain dose of marriage, comfort or social prestige and, therefore, also for a certain portion of truth and untruth. Today, especially in intellectual circles, existential untruth is part of a gentlemen's agreement, never to be mentioned as it is taken absolutely for granted by those involved in the discussion. In discussions and conferences immunising strategies are used that serve to protect the participants from personal consequences. Robert Jungk, an experienced, well-travelled journalist, open-minded to the current global situation, stated his assumption that there are around two thousand people on Earth who are sufficiently committed and free to realise a cultural project of the kind outlined in this book. However, if one could build a model community where the ideas sketched here were realised, it would have a unstoppable multiplying effect. The time is ripe for it, like an over saturated salt solution shortly before its crystallisation.

A DIFFERENT VIEW OF SUFFERING AND ITS CONSEQUENCES

There are certain things that form the culture of a people and their way of life strongly from within. One of these is the way of relating to pain; bodily, emotional, and spiritual pain. A notable characteristic of our present civilisation is that we fear pain and take refuge in therapy. There have been other cultures, for example those of the Native Americans, that were directed towards overcoming the fear of pain. They did not do this in order to raise heroes but to make the human see. A Sioux chief once said:

"My people – there is no growth without pain, nor is there pain without growth. We do not hide from pain, death, or life. The Western people turn away from it; they get their meat wrapped in hygienic packages, cleansed of blood. They try to deny the holiness of the life they destroy when they kill a mosquito. Since we do not even hide from the smallest cry of pain, we will survive."

Life creates the new by destroying the old. The seedling grows by bursting its pod. Every creative person has his ups and downs. Suffering is part of a creative life, since a creative life means overcoming the old, shattering old cages, crossing boundaries. Our age is so complicated and contradictory that the general ideal of a simple and calm life, free from suffering, hardly has any creative meaning. The richer a human is, the more he incorporates the questions, doubts, and contradictions of his time. How could he be without pain? The human is, on all levels, a "becoming" being, anything in this process of becoming not yet solved creates friction, conflict, and disarray. So it cannot be simply a question of getting rid of suffering. We could develop a more conscious, quieter, more active, and more determined way of relating to suffering. Inner healing is a process through which suffering is transformed into consciousness, collectedness and energy.

To a creative person each serious and prolonged suffering probably has a creative meaning, though the meaning may at first be hard to define. Looking back at a later date shows what has changed: a more sensitive, firmer, more incorruptible face, an obvious growth of personality and spiritual substance, a keener critical sense, and a more balanced view of human values.

There are two types of suffering: the creative suffering associated with growth, and the pathological suffering that is associated with paralysis. In creative suffering there is an active life energy and an inner resistance to pain, stemming from an intact sense of identity. Pathological suffering describes the malady of our times: it takes the form of life that has not unfolded but is blocked and twisted from within. At the point where the elementary human drives and growth forces collide with social boundaries, the human splits into a "normal" part that conforms to social customs and an "other" part that keeps seething in the dark, irritating one's daily life with agitating signals. Truly humanising the human world means redeeming the "other" from its repressed existence and integrating it, piece by piece, into everyday life. Anything that the human has not raised to the state of conscious action keeps pulling him down continuously; what we do not truly master, rules us. The repressed part of the human shows us its violated nature through the stranglehold of neuroses and psychosomatic illnesses. But what is this "nature" of the human being?

I am the hermit, crawling out of a cave, with the decision to see the world in which I want to live from now on with a non-judging interest and as it really is. Nothing would strike us as more monstrous, more contradictory, and more incomprehensible than the human being, if we were to no longer prematurely explain or define him with our embryonic powers of understanding. It is only about seeing, not about interpretations and judgements. If we make a cross section through our history to date, or through everything that is happening among people and between people and the rest of nature at this moment on our planet, including that between animals and humans, then there is only one limit to what the innocent eye can behold: the limit of how much horror it can endure seeing.

The human: it is he who built the pyramids, it is he who destroyed cities down to the last child and the last cat, who sang hymns and erected cathedrals, who roasted people with other beliefs over burning coal, and turned those of another race into soap. This human hated out of love and murdered in devotion, preached love of thy neighbour and produced Napalm, loves peace and now prepares its own annihilation.

The inconceivable is demanding an answer. But let us take care not to provide any answers prematurely. We know and accept the correlation between repressed instincts and drives on the one hand and cruelty on the other. The discovery of this correlation is one of the greatest and most hope-instilling feats in the history of our culture. Freud and Reich are the pioneers of a more humane world. But we have reason to suspect that what characterises the human on a psychoanalytical and sexual-economic level may well, at the core, arise from a more fundamental pattern of our existence or our evolution, perhaps even of the universal process of "becoming". The Vikings, though they probably suffered little from the repression of their drives, nevertheless lived lives in which murder and destruction were often a central part. It is this enigmatic tendency towards excess that lurks deep within the human and that has, repeatedly, turned the human being into a monster – or into a saint. *Excess, ecstasy, the crossing of boundaries, the dissolution of the "I" and union with "the other" shines through intoxicating or sacred manner wherever there is a sudden opening. This boundless exaltation of inner energies has been and remains the darkest of motives in human history.* It touches on all forms of horror, love, and religion. What do we know of the human, this germinating being? Measured in evolutionary terms its history has just begun.

Obviously something takes place within us which excites and touches us, through simply perceiving this world beyond the human one. We cannot fathom this reality, we find it is entirely different from what we had expected. There is an aspect of formidability that enters into all that is familiar as soon as we begin to know more. Matter and the immensity of nothingness, into which matter seems to dissolve when viewed microscopically. The living world and the enormous complexity represented in a single cell. Evolution and the gigantic theme represented in the time-span involved. The starry sky and the enormousness of its whole existence, represented in its dimensions. The domesticated cat and the colossal ancestry represented in its age-old eyes as a beast of prey. The formidability of what was done by inconspicuous family-men in Auschwitz.

Behind everything familiar there is always and everywhere the "other", which enters our present from an unknown distance and an unknown past. Its threads of influence meet and unite in the cellular structure of the human and, under the surface, create an explosive mixture of crimes, dreams, and insanity, that influences our experiences, our humanisation and our character at least as much as all conventions of customs and morals have.

Here there is plenty of cause for suffering. The unfathomable appears threatening. Human life seems to be surrounded by an unpredictable horizon of catastrophes, as has always been confirmed when the dam of convention bursts open. We live in this formidable world, are surrounded by its breath and its vibrations. Life that has been shut out threatens us from "beyond", that is, from the repressed regions of consciousness. Every attempt to shield oneself through repression is ultimately directed against life itself. The threat is in no way merely imagination. We are truly emotionally and biologically threatened as long as we try to exclude from life whatever appears to be demonic. The paranoia of our time is pointing out the facts; the ecological and human catastrophe that approaches is a kind of revenge of misconstrued and violated life. The new culture that needs to be created may no longer set up barriers against the invasion of living forces. Its essence is rather the cautious but radical inner opening, the transcending of all barriers, the continuous working on taboos, and the reforming of all life structures. Its psychic and social rudiments should give rise to a vessel that can accommodate the whole human.

All appeals to reason and morals sound like primary school verses when compared to the appalling formidability of the world. The promises provided by past religion and the present promises of therapy are like a narcotic that induces sleep but delivers us into bad dreams. By bringing about new personal experiences, therapy can help us to change track. What then follows is cultural work on ourselves and the circumstances of our daily lives. Our suffering is a signal of a life not lived. Healing consists of recognising, and living, that "unlived" life. To surmount our deeply ingrained restraints, our fears and weariness, our much too cosy humaneness, and our alternative gardens of refuge, we need an experimental milieu in which such a process of change is understood and wanted.

Our consciousness needs to be freed from its everyday limitations and yet remain rooted in everyday life, without mysticism, and without resorting to the substitutes of art, education or religion. What is true in mental-spiritual values must be understood and realised in everyday life. This "must" is not a random one; it is the condition for our humane survival. Religion and art, dreams and insanity, sadism and war, are all areas where our lives unite and collide with the unknown. A central area of collision is sexuality. Compare the repressed, demonic sexual forces that show themselves in fantasies or even in occasional excesses, with the timidity of our actual sexual advances. The genders, in reality raging natural forces, act like soulful sheep. A flood of sensual and creative power has been blocked by a dam of convention, caution and fear. What remains is a trickle that cannot quench the real thirst. Many hardly notice it any more, for they have learned to protect themselves from Eros – through sexuality for example, or what we commonly call sexuality.

Sexuality is only one example, though it may be the most fateful one. Anger has suffered a similar fate – the great biological anger that is brought forth by the unbridled forces of life itself within us, when life is trampled down or obstructed. So has curiosity, the passion for adventure, and any other form that our natural need for expansion takes. We have become too fastidious, too tame, too small. We are constantly holding back. That is the origin of our perpetual suffering with all its psychosomatic symptoms such as migraine, impotence, depression, and even cancer. It is the continuous depression of a life imprisoned in a ghetto, incessantly bombarded with impulses from the "other" world, the side representing the life possible beyond boundaries. This whole subterranean abyss between the potential that we have as cosmic beings and our guarded forms of interaction underlies the permanent irritation that is fashionably termed "conflict between mind and body" or "not being able to let oneself go". In the face of the actual theme we are talking about, the modern consciousness of suffering, coloured by psychology and therapy, is like an endless emotional chit-chat. It has become almost chic to speak of one's own fears, back pain, or sexual troubles. Talking about suppressed emotions has developed into the most subtle bastion against a real understanding of the situation and real desire for change. The defamation of the intellect, so common today, has done the rest by switching-off the authority capable of giving us an overview: the head.

The suffering of our modern society is the self-paralysis of elementary life processes at the levels of sensuality, mind, spirit and soul. From a larger perspective this paralysis is the result of a falsely programmed system that creates and sustains the forms of our everyday life and social graces. They are too small; the human phenomenon does not fit into these forms. Beneath the layer of social behaviour is an immeasurable potential of vital, erotic, spiritual, and expansive energy that so far has found no space for free human and societal development. Behind the emotional and psychosomatic illnesses of our age lies the basic pattern of fear resulting from constriction. The old forms of convention and custom are too constricted, as are the old forms of marriage and the nuclear family, the forms of highly specialised division of labour, the forms in which new theories are developed and science is practised, and the forms of political practice in the subcultures of the left and the alternative movements. The "other" in life and the truly creative, whose objectives and source of nourishment is the new, the unknown, the highly tensioned, and the contradictory, does not fit inside of them. The principle of constriction and fear operates within. Therefore, in order to integrate ourselves into the Living we need an *inner expansion* of our lives to overcome our emotional and bodily armour, to open the entire organism and centre our consciousness at a point of ever deepening identity. Inner centring and free communication outwards are two sides of the same process of expansion that connects us to the living world. Everything that follows depends on this process. To set this in motion we need societal and individual strategies for overcoming fears and constrictions. That is the key statement for building a new culture. It includes everything that can be done to create inner expansion: a new system for professional life, new architecture, new forms of living together and of raising children, new forms of love and sexuality, new methods in medicine, new possibilities in art and self-exploration. The creation of a culture in alignment with the laws of life, first requires the creation of a new social and psychic space where people can again learn to live and breathe freely. From such places the most dependable healing power against fear and hatred would arise – love.

3 THE TRUTH OF THE LIVING

FEAR AS A BIOLOGICAL DISEASE OF CULTURE

The alienation felt by people today is, at its deepest level, a biological one. Our cultural disease consists of the loss of the Living and of the sense for the Living. This can be seen in every modern hospital, including the strange types of medicine practised there. Concerning the questions of life modern man has become fatally stupid – fatal in the literal sense. In West Germany someone dies of cancer every four minutes and the number of "psychosomatics" is enormous. Our poisoned and congested environment and the overly specialised working systems with their fragmented detailed work cause neurotic and bodily tensions and bioenergetic blockages in a large part of the population. These factors, combined with an overall loss of direct human communication and the desolation of human relations, leaves contemporary lives in a state of inner constriction that deprives them of calm, depth, and space for the fundamental experience of the Living. *Constriction = fear.* The biological disease of culture of our civilisation is fear, fear as result of bioenergetic and soul blockages. And fear is love's adversary.

We need to fully see and understand the extent to which "humaneness", understanding, and tolerance, gentleness and consideration, care and sympathy, are in truth the result of fear. If one has ever been a clear witness of the exchanges taking place in the sentimental theatre of words and emotions we all take part in, then one has an inkling of what it means to build a community without fears and lies. This is the most fundamental change that we can make. We do not usually notice the fear because our commonly accepted moral agreements, cultured conversations, ideologies, and habits constructed of it. Fear is "bound" within the system that we take for granted in our daily lives. It is the major psychic ferment emotional catalyst of our entire culture. Most people cannot even imagine what it would mean to love without fear, as what they term "love" is connected so plainly to the fear of losing someone, sexual fears, fear of authority, of rejection, of being alone, of betrayal, that the absurdity of the situation is no longer recognised. Only the results become visible as jealousy, illness, depression, and broken relationships. Love without fear is without doubt the opposite of what is termed love in our culture.

The truth of these interrelations belong to the most incredible aspect that our time has to offer. Anyone who does not see this, at least in part, should close this book at once, because to him what follows will seem pure nonsense.

LIFE RESEARCH AS THE SCIENCE OF THE FUTURE

Building a culture based on the laws of life requires a deepening understanding of elementary life processes. Elementary processes are universal processes that are connected with all levels of life, from the simplest plant realms to the soul regions of the human. They are connected with germination, growth, and development, with rise and fall and re-arising.

This drama of life is reflected in the inner situation of the human – in the process of maturation, in the dynamics of drives, longings, and in the structure of conflicts. The same laws of the Living operate both in the inner nature of humans and in outer nature. Cultural work has to be oriented towards these laws if we wish to regain our sense of biological, energetic, and mental-spiritual identity.

Life research is a science of the future, perhaps even the science of the future. What we can say about the functioning and structures of the living world has a temporary character. But even the little we can see is contradictory to our habitual ways of thinking, the ways we formulate scientific theories, or our concepts of "objectivity", how we see our lives, how we structure the conditions of our lives, our morals, traditional medicine, pedagogy, or architecture. We can, with certainty, say that a culture and society oriented towards the laws of life will develop along principles that are diametrically opposed to those developed by our present-day culture and society. Life research relates to common science as sight relates to the anatomical structure of the eye.

The natural order of the living realm is fundamentally different from man-made systems (technical, juristic, governmental, military). It is more complex, unified, open, and less susceptible to dysfunction. Most importantly, it contains the impetus for development. Its functional principles are rhythm, communication, resonance, lightness, radiation, tension, pulsation, and polarity. An incredible precision is achieved through imprecise methods, such as touching, feeling, oscillation, interference, etc. Not until we acquire a new system of thinking will we be able to understand the extraordinary functional principles on which life is based, step by step. We need a new philosophy of the Living, and a new kind of biology to shed light on the magic of life, its "occult" laws and psi-forces, its mysterious processes of communication, its principles of organisation and growth, its logic that governs functions and developments in nature, its fantastic technology, its energetic processes of communication and resonance, and its pulsating, radiating, metamorphosing nature – this biology as the comprehensive science of the Living will be a central theme for the new cultural era.

Research has to do with cognition. Seeking insight is no human luxury, for it lies at the heart of evolution, and evolution is a process of growing cognitive capacity. Our sense organs are rudimentary organs of cognition. Perhaps the greatest and most accessible secret of the living world is its endowment with sensory organs. They serve perception and communication. Obviously, the guideline for evolution was the development of increasingly sensitive and far-reaching organs of contact for perception and communication. Perception and communication seem somehow to be fundamental to the living world. Why? What secret of providence was in this way meant to stay readable in life? Sensuality is perception and contact. It is, therefore, associated with cognition. The evolution of life has brought forth different forms of cognition, for example, sensory perception, intuition and the intellect. The intellect is the youngest and most inexperienced child in the evolution of cognitive life. The faculties of cognition through sensory perception and intuition must be intact for the intellect to function in a meaningful way in terms of evolution. Since these faculties are no longer intact life research requires a larger setting where sensuality and intuition may regenerate. An organism that is awake, with open pores, through which energy can freely flow, is a single sensory organ to experience the world. This experience is collected, reflected upon, and steered in the centre of consciousness. Cognition of the Living means opening the senses and awakening this centre.

CONTACT AND TRUTH

Cognition has everything to do with truth. But what is truth? Before it can be discovered and formulated intellectually, it already exists in an entirely different form. "Truth", says Wilhelm Reich, "is immediate, full contact between life that perceives and life that is being perceived."

This is one of the most profound statements in the theory of cognition. This means that truth is not primarily a matter of thinking but of *contact*. However, contact is a sensual and basic bioenergetic function of the living world as all living organisms are in contact with others. Thus truth is also – as long as the contact has not been broken – a basic function of a healthy organism. Most words of our language get their meanings primarily not from an intellectual consensus but rather from immediate contact. A child learns the meaning of the word "bitter" by hearing the word and associating it with the relevant facial expression. It is experienced through contact and understood intuitively. So the world of contact precedes the intellectual world. When the contact is disturbed the intellectual world is also disturbed, since *it has lost its biological, emotional and energetic preconditions.* The era of traditional, natural science and the corresponding structuring of society has been an era of increased loss of contact. Full, immediate contact with the living world is severely distorted through the barriers of fear, embarrassment and revulsion – especially to areas related to the excretory functions and sexuality. Truth is realised through overcoming such ingrained barriers. Life research implies working on taboos, for hidden knowledge manifests in the act of crossing such boundaries. The terms ecstasy, excess, and transcendence signify this element in their original meaning. Surrender would be a word for the according way of living.

THE TECHNOLOGICAL WONDERLAND OF THE LIVING

What is nature and what is life?
If I observe – in full consciousness – the shoot of a plant emerging through the earth in a flower pot, then at once I am faced with the entire enigma. Who am I, who is observing and thinking? This question points to another facet of the same enigma. It is a strange characteristic of the human mind that it can overlook this ever-present mystery simply through force of habit.

Something that is outwardly so familiar to us that we scarcely notice it contains within it a world that may one day answer all our questions. It is a world of structures and functions that have grown organically, with nature allowing millions, even billions of years for its development.

We step into an organisational and technical wonderland. The form of a seashell on the shore, the static system of a hollow bone, the structure and dynamics of a vortex, an animal's sense of orientation, or the system of information exchange within a cell – all this is amazing and exceptionally interesting. Yet it all gives us insight into only the outer façade.

What kind of energy technology could appear on the horizon of a new culture if we could get an inner view of the functioning principles that govern a blade of grass as it breaks through a layer of asphalt. What possibilities would occur from the soft emergence of power if we could understand the principle of resonance that so obviously operates everywhere in the living realm? The architecture of a corn-stalk, the construction of a spider web and the grasping technology in the claws of a dormouse – these are truly miracles of stability and "efficiency"!

The technological approach to these things is fascinating, but go to the heart of the matter. Once again the human mind, asking for measurements, proportions and chemical composition in order to reconstruct things, bypasses the essence of things, namely that these works of wonder did not arise as the means to an end, but rather as the result of a life activity on the part of the living being concerned. Behind all the biological perfection there lies a certain state of being in the Living, a constant feedback and attunement with the structures of the universe and surrounding nature. The biological message that lies here extends far beyond the technological one; it is eminently existential in nature. The observable structures and functions are the empirical manifestations of a mystery. This mystery lies in the way of life and the universal "method of existence" of the Living.

THE HUMAN IN THE ENTIRE ORGANISM OF NATURE
Every living entity has an outer and inner aspect. What is within is, no matter how rudimentary, of a psychic nature. The psychosphere is part of the biosphere. The processes of life are directed from within. Evolution was certainly also a game of chance, but it would never have come about if there were not, in the essence of every living being, a germinating centre that could choose and grasp the results of chance.

That life actually has an inner aspect is beyond all doubt since evolution has produced a being that is capable of looking within – the human being. In the entire organism of nature the human is the eye with which life sees itself. When he looks within, he is an authentic witness to the inner aspect of life. What he sees and experiences within is the most immediate form of life in its fullness (though it is often contorted and repressed). The inner impetus of the evolution of life, going from one form of life to the next, producing the whole spectrum of animalistic instincts and drives, reveals itself in the as yet most developed form, in humans – as desire, longing, will, emotion, and "soul". This constitutes the inner continuity in the evolution of the living world. A science of biology that continues to concentrate on only the outer forms, without expanding its possibilities to experience the inner aspect, no longer has any historical meaning. The same is true for medicine and architecture. Life does not let itself be divided. The Living that operates and perceives, questions and researches in the human, belongs together with the Living that moves plants and animals from within.

All these interconnections have methodological consequences. Since all that lives has an inner aspect, the new science will be dependent upon the power of recognition through *intuition*. This does not mean that the objectifying and analysing power of intellectual thought is discarded. The battle between the two is an antiquated episode in the bizarre and confusing human history of ideas.

Perceiving and experiencing life is largely a question of capability for inner experience. The meandering flow of a river, the singing of a bird, and the survival of a deer in the most severe winter can somehow be described and explained in terms of their physical aspects, but they can only be truly understood from within. However, the capability for inner experience is to a large extent blocked in our culture. The barriers consist of the obstacles that, in the name of morals, human dignity, and a misconceived notion of humaneness, have been created against the "animal in man". Many elementary biological, vegetative, and animalistic zones of experience are therefore unreachable by our human consciousness. Most grown-ups can no longer cry fully, be fully enraged, love fully, relax fully, or breathe fully. Through this something has been lost in the human substance, in his direct experiential world that is necessary for a true connection and union with the Living. The human has disconnected himself, through false programming, from elementary life energies and life processes and has ideologically supported this separation from its own source by creating the concept of "objective science". This objective science has observed and analysed only the outer aspect of the world. The deeper reason for this in no way lies in the essence of science, but in the *way of living* of the scientists and the culture they belong to. They have observed the world externally *because they truly stand outside of it.* The old magical and mystical connectedness with nature and cosmos was gone. Further barriers to communication and emotional blocks have closed off the processes of open communication and flow between Human and World. In this way objective science has become the hallmark of disconnected humans and their era.

FUNCTIONAL PRINCIPLES OF THE LIVING
a) Unintentional and Effortless

A dog that has lost his master when out walking in an unknown forest has no intent or plan for getting back home. It follows its instincts at each moment. With open senses, sniffing at and urinating on the curiosities along its path, it goes its wandering way and presently arrives back home well before its worried master.

Not only the dog, but every living being, all life processes in plants and animals, and all systems in nature function without effort and without intent. Even the most powerful movements of a panther happen without effort. That is the secret of their power and beauty. The Zen culture of the East follows the same principles in the art of archery and swordsmanship (of the Samurai type): the highest beauty and perfection without intent or exertion. It is the principle of the Centre ("Hara"). One who is at rest in his centre has cosmic powers at his disposal – like a blade of grass, a tree, or an animal. The artistry of life, when it is not disrupted, is that this calm is not lost, even in movement.

The capacity for having an intention and pursuing a goal is a relatively recent development in evolution that has only found its original expression since the emergence of the human. This capacity is a result of evolution; it has thus been added to the abilities and potency of the Living. We should not, therefore, try to establish a way of living without goal and intent, mimicking a lack of will. The goals we set will only be meaningful in terms of a life-oriented culture if we see and understand a way of existing without intent and effort as a universal principle of the living world. We must – in the sense of humane progress – again become capable of participating in this mode of being.

b) Pulsation and Peristalsis
In a true orgasm the human is like a convulsing mass of plasma. This "undignified" fact constitutes one of the deeper reasons why sexuality became taboo. The opposite should really have occurred. The fact that in the orgasmic convulsion we are dealing with a primary biological function of great importance that the human has in common with all living beings, down to the jellyfish and the amoeba should be reason enough to retain and protect this function at all cost. Whenever a living being moves independently, without being subjected to outer interference, one of the principles according to which it moves is that of pulsation: a rhythmic sequence of expansion and contraction, charge and discharge, tension and release. The organs of our autonomic system, such as the heart, lungs and stomach still show a fairly pure form of this rhythm (not superimposed by other rhythms). Worms move in this peristaltic way. Pulsation and peristalsis express a basic functional principle of the Living.

Modern man, due to his urge-adverse cultural tradition in a psychic and bioenergetic overall structure (body-armour), has largely lost contact with his own elementary biological functions. But if this body-armouring is dissolved (through a strong emotional experience or body therapy), the body immediately surrenders to these long suppressed basic functions. Convulsions tend to develop automatically, centred in the stomach area – convulsions that soon envelop the whole organism. This process is often accompanied by screams, sobs, and a liberating cry. The human is again at one with himself, with his basic biological processes, and thereby automatically also with his soul identity.

It was foremost through the research of Wilhelm Reich that these simple truths were brought to light and placed in the context they belong, of a fundamentally new sexual and emotional order of human culture. The full energetic sexual orgasm is a key function of life. One can only imagine how far-reaching and deep a change of attitude and perspective must be achieved today before the simplest interconnections in life can be seen again. Life research contains enough implications for cognitive theory to confound the representatives of classical science. For example, there is a strong correlation between the ability to research life and the ability to achieve orgasmic convulsions, sexual surrender, and general life potency.

c) The Contradictory
To justify its name, life research needs a new way of thinking. The concept of a "life-oriented" culture should not be misunderstood as a mechanical copying of the laws of life. The patterns and processes of the living world can neither be understood nor applied in a mechanistic fashion because of the complex and contradictory nature of the living world. There is a phenomenon known in therapeutic body work that can be verified by trying it on oneself. Pressure on a sensitive part of the skin causes discomfort, but a small change in one's inner attitude can trigger pleasure. These things are ambivalent. Even a small change in mood or perspective, an unnoticeable crossing of an invisible boundary, causes a phenomenon to be appear as its opposite. Definitive statements such as "like this and only like this" or "either-or" become inapplicable when dealing with the living world. Nature never follows a straight line. The deeper we delve into the living world the more we discover the intimate belonging of every truth to a counter-truth, every principle to a counter-principle, every thesis to an antithesis. Life seems to work in a fundamentally antithetical and polar manner. This universal duality and ambivalence of the living world also runs through the deeper levels of human consciousness surfacing, for example, in the "double meanings of primal words" or in the "twin faces of archetypes". This phenomenon has been well documented by depth psychology and

by the study of mythology. The difficulty in understanding and conceptually grasping the Living arises, among other things, because the two opposites form a functional unity. Examples of this unity are:
Construction – Destruction
Law – Spontaneity
Order – Chaos
Complexity – Simplicity
Tension – Relaxation
Flowing – Forming
Movement – Stillness, and so on.

Growth is flow *and* form. Creation is freedom *and* necessity. Freedom is spontaneity *and* regularity. Development is motion *and* stillness. Evolution is determination and (growing) freedom from determination. To every cardinal thesis there is a cardinal antithesis. Only both together yield the truth about the mysterious process we call life. Compare that to the whole register of dogmas entailed in our morals and values and the common mode of judgement! Would it not be worthwhile to think analogously and apply this *phenomenon of paradox* to human and social life? The individual's development results from an inward centring combined with an outward expansion, contraction of the "I" and "dissolving" of the "I". A living community incorporates both spontaneity and organisation, organic growth and design, centralisation and decentralisation, hierarchy and democracy. Both the ecological outer space and spiritual inner space of a community need to be cultivated. It is desirable to be grounded in the reality of daily life and in the growing certainty of the transcendental world. (The issue of antithesis is raised and discussed further in the postscript concerning tradition.)

d) The Open

All life is open: it is in a constant communication and in a constant state of becoming. Every living being, every biological unit (for example a biotope), is in terms of modern biology, an "open system". This biological openness of a system has two aspects: there is not only the kind of balanced flux where the system is in a constant circuit of exchange with its surroundings. Another aspect is involved, the aspect that the French researcher Edgar Morin, known for his attempt to create a more deeply rooted biology, alluded to at the end of his book *Le paradigme perdu: la nature humaine* ("The Lost Paradigm: Human Nature"):

"One becomes stronger when one forever wrenches loose from the magic word that explains everything, from the litany that supposedly solves everything. And one gets stronger if one sees the world, life, humans, knowledge, and action as open systems. The opening – a breach into the unknown and into nothingness – is at the same time the thirsty and hungry mouth through which our spirit and our life desire to breathe, drink, eat and unite in love."

The biological quality of openness has something to do with the basic fact that every living thing is in a process of development, and this development obviously has an "inner" aspect, a directing, invisible potency – we speak today of "self-organisation" – that seeks to enlarge the scope for spontaneity, freedom and openness in the course of evolution (Teilhard de Chardin based his theory of evolution upon this).

No matter how we wish to interpret the openness of biological systems, it sets standards for building culture and society on the laws of life. One of the most important qualities of a living cultural concept is openness to change, self-correction, and development. The translation of the principles of "open systems" to human reality, as Popper's Critical Rationalism for Science and Bloch's *Experimentum Mundi* aspired to do – not

ually overcoming all rigid ideo-
tual structures – is today a pre-
iral concept. Humaneness is ac-
:tory and the open.

most accurate relationship to an
ist it. A blind person reads Braille
is fingertips. Truths often sink in
ig hinted at. Dogs get acquainted
two leaves are identical. Biologi-
:d by "static noise". A stream does
ut in meanders. The shortest path
proverb. Life often achieves incre-
n bees build honeycombs, animals
ls transmit information to other
achieved through the "imprecise"
ng, oscillating. *Precision lies not in aiming, but in contact with the continuous feedback of the signals sent out.* Once again it is more a particular mode of being than a technical method that makes the function possible. It is similar to trance states where one remembers things that could never be recollected through conscious effort.

A NEW MENTAL-SPIRITUAL ATTITUDE

If I wish to experience the fragrance of a flower I smell it intently but very lightly. The method used depends on the type of smell involved and the way the sense organs function. Life research is also of this nature. We are, with regards to science, in the transitory phase of a fundamental paradigm shift, and this means a change in the mode of experience. In order to study nature in a meaningful way we must acquire a different overall outlook that will itself unfold the new questions, methods and concepts for the work that is to be done. For life research we need a comprehensive philosophy of the living world before we can devote ourselves to a meaningful kind of analysis and judgement. The manifestations of the Living are too diverse to be sought and grasped without a tactical scheme. But at the same time they are too profound, paradoxical, dialectic, and holistic in their functioning to be understood within the model of traditional biology. Life research encompasses, among other areas, biology, ecology, bioenergetics, psychosomatics, and depth psychology, uniting them in the framework of a universal biophilosophy. The emerging methodology is no longer of the hard scholastic and definitive kind, as is modern science, but softer, more open, and flowing. The phenomena of paradox and system openness, as well as our inferior standard of knowledge in life research, compel us to take an intellectual attitude that fundamentally affirms doubt through its ability to remain in tense suspension between two opposites without immediately needing to resolve the issue (in order to attain a psychic relief). Ultimately, difficulties are overcome through ease.

4 BIOLOGICAL HUMANISM AS THE FRAMEWORK FOR A NEW CULTURE

THE CONCEPT OF BIOLOGICAL HUMANISM

Throughout history attempts to improve the world with morality and religion and to conquer man's savagery by appeals to reason and conscience have failed. Men are humane to the degree that they recognise and fulfil their bodily, emotional, and spiritual needs in a *social* way (that is, in living together with others). The philosophy of the new culture does not appeal to any morals but to the deepest and most conscious "egoism". The necessary cultural change will not come through appeals or sacrifices but by changing our life practice, our way of working, our human contacts, our sexuality, and so forth, in a way that *fulfils our needs*. A new cultural concept can only be realistic to the extent that it presents a model for a better and truer self-realisation and fulfils a greater number of basic needs.

A culture based on true needs would, of itself, have a great ecological advantage. A community with loving communication, fulfilled sexuality, and creative work would no longer be dependent on vicarious satisfaction through the products of industrial society, as is the case today. It could, therefore, develop a new *consumer model* that would avoid wasting goods, energy, and other resources, thus saving the environment from destruction. The new consumer model needed by the ecological society of the future, is first and foremost, a new model of human self-realisation.

Humanism is a spiritual impulse towards liberation that entered history in the early Renaissance and has persisted through the eras of humanism, enlightenment, science, Marxism, and psychoanalysis. Slowly man begins to see the world as something to be observed, grasped, analysed, and changed.

He frees himself from the old order based on authority, religion, and society, and on taking the reality of a situation for granted. By constantly using his own reason to enter into new realms, he creates the necessity to re-organise and re-form his world. This is the essence of the genuine humanistic tradition: to wrest control from formidable old power structures in order that the human steps into self-management and responsibility. On this *historic path to autonomy,* humanity can tolerate no pre-ordained limits and no moral ones, for it follows evolutionary laws of development. Morals themselves become the object of analysis and change. Nietzsche, so far, dealt most consistently with these matters. Through Copernicus the power of thought entered into the religious world-view; through Marx it broke into the bourgeoisie; through Nietzsche it broke into the moral world-view; and through Freud it entered into sexuality. Every new breakthrough opened a new dimension in human social life that needed to be dealt with, assimilated, and integrated.

The dimension that now needs to be discovered and integrated is the dimension of life itself with its special principles of holistic and ecstatic ways of functioning (see previous chapter). The forgotten biological basis in which all control mechanisms are embedded must be laid open. The organic system of body and soul with its drives and emotions (that has an evolutionary and therefore historically changeable character) must regain its natural ways of functioning. We can only break the alienation of our age by reconnecting the human forms of action and development to the universal processes of the living world.

Biological humanism strives for the maximum integration of the human social world into the overall realm of living nature (I do not know whether there is also "dead" nature or whether what is called dead is in reality only a special case of the living world). Such an integration will realise the ideas of "organic environmental design", like that developed by Hugo Kükelhaus, and will also achieve a fundamental reorientation of man in relation to his own nature. The nature of man consists not only of his anatomy and his physiological processes – here we share Teilhard de Chardin's concepts of nature – but also of all his emotions, drives, instincts, and energies.

One of the central inner drives of the human is the sexual drive. In this area there is a common denominator running through almost all cultures, religions, morals, philosophies, and political ideologies of the Occidental world, and that is the secret or admitted capitulation in front of Eros. Show an upstanding theorist a picture of a well-proportioned female with an inviting cleavage and he grows pale. If the destiny of the Earth and our culture is to be put in the hands of adults then these must be people who have free access to their erotic powers without suppression. This is the only way that the living world can be "controlled".

The same evolutionary lines of development, instincts, and drives that are present in the animal realm come together in humans at a higher level. Becoming human is, in its most far-reaching sense, a progressive spiritualisation and sublimation of all animalistic forces into a human culture. This process of sublimating has not yet succeeded because instead of accepting, cultivating and refining his animalistic strengths, man has tried to suppress and ignore them. We can perhaps see it as the "basic error" in the course of the history of consciousness that the process of cultivating the human animal was carried out as a battle of the spirit *against* the "beast in man" instead of through the union and reconciliation of the two. It was a fight against nature itself, which naturally could not be won. Instead of a sublimation of the *whole* human with all his drives, a dangerous split occurred, into an official and a repressed part that has so far obstinately resisted all attempts at humanisation. *In this psychological ambiguity of the human lies the principal illness of our age.* The conscious exposure and reintegration of repressed material and the "acceptance of the shadow", which C.G. Jung formulated as a therapeutic principle, must be taken from the therapeutic level to the social and cultural levels, if man is to become whole again.

The human must become a conscious fellow creature on Earth, if not, he will senselessly destroy it. He can only achieve this if he accepts the authority of life and submits to it. Humanisation is the humanifying of the Earth, the penetration of the human into deeper and higher realms. But this penetration means mental-spiritual transformation resulting in a non-imperialistic domination. There is a law of the living world that only allows spiritual expansion through spiritual transformation. We cannot control natural forces by fighting and conquering them – through that their unpredictable nature runs out of control and leads to earthquakes and floods, also on the soul level. Seen in this light, the history of our culture has largely been a puppet theatre, directed by the strings of the repressed, "conquered" natural forces and life energies. If one seeks control in the living world one must unite with it, get to know its rules, and follow them. It is a totally new kind of control, no longer dependent on contest and suppression, but on ecological harmony.

THREE STEPS TOWARDS A REALISTIC HUMANISM

The emergence of a new culture contains in some sense a theme of political theology. A political or rather a societal concept needs to be developed that, in its depth and existential meaning for the individual, is equivalent to the religious ideas of the past. What were once steps towards inner individual transformation are now steps towards a metamorphosis of the social fabric in which we live. In this social fabric – in our work, our social institutions, and our human relations –

we must one day be able to occupy and truly know ourselves to such an extent that we need no other comfort and no other home outside it.

At some point in the development of man, the human mind deemed it necessary to go against the body and its sensual needs. Cultural development thereby took a path that led the human away from the entire organism of nature, to which he wholly belongs. Since then, history has resembled a dance around an unknown centre.

Religions have tried to uphold a vision of a better afterlife to compensate for earthly misery. Salvation lay in freeing the soul already here on Earth as much as possible from the physical world, for it was identical with sin, a prison, or maya. The goal was thus to conquer the body, to conquer sensuality, and to conquer earthly misery through mental-spiritual exercise. We find this fundamental idea in all the religious leaders of the past, from Buddha to Aurobindo, Plato to Rudolf Steiner, and from St. Paul to Pope John II. The idea was by no means a false one. Since the soul is truly an independent entity that can actually free itself from the body (as has been done, for example, by old cults of initiation, in religious ecstasy, in LSD-experiments, in peak experiences or in near-death situations), this healing concept was realistic. But it led the healing interest away from everyday life on Earth and away from earthly human longing. The atrocities on Earth continued unabated.

Next to the religious impulse towards liberation, we find the political one, a much later phenomenon that is still today in its early stages. It has so far found its most unequivocal philosophical formulation in Marxism. Marx's epoch-making idea was to annul (and redeem) religious ideas of liberation through political practice (the class struggle). Salvation was no longer to be erected in heaven, but in the most materialistic point in the physical world: in material production. "The criticism of religion" said Marx, "ends with the teaching that man is the highest being for mankind, that is, with the categorical imperative to overthrow all conditions in which the human is a degraded, enslaved, abandoned, and contemptible being."

By creating a new order for human labour without class domination and alienation man was expected to find his centre and home in his everyday social practice. This mutational leap in the history of ideas was the most revolutionary feat so far achieved by the human Prometheus. It acts as a signpost from which there is no road back. But Marxism was not yet capable of thinking and formulating its idea of political self-liberation at a deep enough level. Its political-economic theories did not truly offer a full equivalent to the religious ideas of salvation: the human had not yet been fathomed deeply enough, his alienation and ultimate longing not yet understood deeply enough.

The next fundamental impulse towards a secularisation of salvation, as profound and as worldly as the Marxist approach but arising from an entirely different point, came through psychoanalysis (we leave Nietzsche aside, who is not so easy to fit in here, and whose work had almost no social impact, because a discreet understanding of his "heroic philosophy" will probably be grasped by later generations). Owing to its authentic humane motive, psychoanalysis was first of all an act of honesty. The puritan Sigmund Freud recognised in himself the overkill of sexual impulses present in the hypocritical culture of the Victorian era. He immediately saw the cultural universality of this situation. He noticed that here, in the libidinous realm, matters of happiness and misery were determined in an area that lay entirely outside official consciousness. He thereby pulled the question of salvation from the afterlife into the "basest" aspects of life on Earth, namely into the domain of sexuality. But, as with Marx's work, sexuality turned out to be in a condition of utmost misery and perversion, as it had for so long led a repressed, insulted, exploited, and hypocritical existence. Freud recognised that the moralistic sexual barriers and sexual structure of the family led adults to live in deep captivity of the soul in a world of subconscious drives and fears, constructed from projections, fixations, and unfulfilled fantasies. He knew that this psychic underground would have to be redressed, if man ever wanted to be free.

Freud's discoveries could have contained the seeds for a prodigious cultural revolution, had he not stopped it himself through his faint-hearted theory of culture and sublimation. In the struggle between needs and society, he finally came down on the side of society, presumably to save his societal position. We are entitled to view that as a barrier of his time and to pursue those unfinished truths beyond this barrier.

The next great pioneer who drew back the veil still further was Freud's successor, Wilhelm Reich. An unusual path of discovery led Reich to realise the identity of sexual energy and universal life energy. In the sexual orgasm he found the prototype and the key to an understanding of fundamental biological functions in all body tissue. In processes such as pulsation, peristalsis, tension and release, charge and discharge, and contraction and expansion, he saw the fundamental activity and functions of life energy itself. These modes of functioning are of a universal nature, that is, they are a part of the universal order of life. But in our culture's human they are considerably disturbed through inner blocks and congestion, obstructions caused by society and morality. Reich termed this "body armour".

This discovery of a universal order of life in the dynamic realm of drives and urges made way for a new vision of liberation. It consisted of a conscious reunion of the human with his most elementary functions of life. The possibility of salvation that Reich found here he called simply health. If the fundamental biological functions can flow freely then the organism, including its aspects of soul and spirit, is connected with the universal order of life and is healthy at its core. But if they are blocked and disturbed, then the organism is disconnected from the universal order of life and is sick at its core. Correspondingly, a society in which the biological currents of energy can flow freely is healthy at its core; a society where they are blocked is sick.

To base healing on the free flow of life energies in the human organism – would that be too one-sided, too narrow, too "biological" a concept? Perhaps. But let us never forget that "biological" does not refer only to what the mechanistic view of nature in the materialistic era has limited it to. In the unsolvable context of Bios and Psyche, life energies are also always of a soul and spiritual nature. Correspondingly, the mode of experiencing the world that spontaneously arises in a fearless and freely flowing organism is of a specific soul and spiritual nature. The world becomes alive. The landscape that I see is no longer purely an image, it is part of creation. One realises that being alive means taking part in creation. It is like an elementary encounter with the world. It leads to new and more intense perceptions, of sight, touch, taste, and smell, a new way of walking and of putting one's foot on the ground. One suddenly understands animals, their elasticity and calm, their way of pointing their ears, and the power in their readiness to leap. The organism becomes impressively strong, light, and transparent, almost musical. In this experience and mode of being there is an element of animal vitality, of soft power, and also an element of intensity and celebration that points towards a new sensual and vital kind of sacred perception. It is the religiosity of universal love that now flows by itself from its biological sources. Reich's descriptions show that he knew this state. To him it was simply the autonomous functioning of life in the unarmoured human.

Reich's advances into the realm of life were a pioneering feat that cannot be overlooked if we today want to lay a realistic foundation for a new culture. Marx's great political thought, to cast off *all* conditions through which the human is demeaned, could now be thought through radically to its conclusion. *All* conditions means the working conditions *and* the psychic-energetic-biological conditions in emotional and sexual human relations. A remodelling is needed both in the organisation our working life and in the organisation of our love life! The entire libidinous and intimate emotional texture of human society must be able to develop anew without restriction and prohibition, without fear and compulsion towards emotional lying. The ecological movement was the first political group to make life itself and the protection of life its main political theme. In this context Reich's thoughts need to be updated. Today it is not possible to realise an ecological humanism without taking into consideration bioenergetic and sexual-psychological interrelations.

AN ECOLOGICAL RELATIONSHIP TO ALL LIVING BEINGS

An ecological relationship to animals, plants, rivers, etc., is an indigenous relationship with nature. The inner interconnectedness of all sentient beings in the biosphere is of a soul nature. Every living being is a soul-being; if it is not a soul-being it does not live. The ecology of living beings, therefore, also encompasses their soul relationship to each other. The universe of the Living is a hierarchy of ensouled and conscious beings, all communicating with one another. All that squeaks and scurries, all that crawls and creeps is an expression of living soul, each at its own level of consciousness. Every level of consciousness has its special way of being in contact with other beings and at the same time represents a special attempt to find a solution for the issue of life. Every level is a piece of evolution, a piece of universal research and curiosity, an organ of the entire organism, and a chord in the full score. Animals and plants are universal beings. Their highly meditative way of existing gives their souls and bodies capacities that in the human realm we find only in the "insane", the saint, and the Indian yogi; for example, the capacity for deep meditative states, concentration and presence, the highest degree of tension and relaxation, the most elegant deployment of power without effort. It allows for incredible feats of orientation, the utmost composure in the face of cold and pain, and a trance-like shutting down of the whole organism. These are all *animalistic capabilities*. If the lab assistants of scientific nonsense working in the torture chambers of modern science, cutting up living and conscious laboratory animals, had even the faintest notion of what they were doing they would at once break down in a sea of tears. This sea of tears is there in any case, invisible and mostly inaudible, in all creatures below the human. Their prospects today are hopeless since they cannot communicate with the armoured, closed, and mechanised human animal. On the whole, children suffer the same fate, as do women and all adults, insofar as they have remained animals and children within. If the life that has been neglected, trampled on, tortured, ignored and scientifically disavowed had a voice, the Earth would be a single scream.

Of course Francis of Assisi could communicate with birds. Every native American could, for living beings were made to communicate with each other. Wherever there is no communication an essential channel has been blocked and a stream of life cut off; in other words, there is a defect in the functional fabric of the biosphere. Communication is a biological, bioenergetic, and psychic process, that as a rule occurs naturally without words. The results of research with dolphins, the research reports of the "secret life of plants", and every unprejudiced concentrated observation of animals show that we are dealing with living beings whose existence and ways of reacting are quite similar to those in human life. The human can, if he is internally prepared to allow it, confirm and understand this. That's just it! In the entire fabric of life the human is the only one who is in a position to *understand* other beings. But instead he usually experiences only fear, disgust or indifference. The simple thought that a rat is an animated, needy, curious, and hungry being could, for a moment, free him of his grotesque fear. As is the case between humans, fear is the real obstacle preventing communication between human and animal. This fear, however, is not usually experienced as such, since man in his absurd blindness no longer even recognises that there is something one could communicate with. A *realised* biological humanism is a world in which the human would naturally perceive his fellow creatures as ensouled beings, he would respect and, whenever possible, aid them. It would be a world in which each day might begin with humans directing their loving attention in deed and thought towards their growing community with all living beings. We can be sure that we would then get more bountiful growth and healthier development, for love is no mere sentimentality, it is a biological force.

This realistic vision will only be realised when the human has shed his inner burdens and freed himself from his present condition in which blocks, anger and senseless activities have left him unaware of the subtle things happening around him. The fate of fish, birds, and mammals will not permanently improve until the human has carried out his own cultural revolution, accepting and making a deep and complete reconciliation with the "animal" within. Any other attempt would be of a moralistic kind and, therefore, unrealistic.

THE IDEA OF SCIENCE

Advancing towards the living world, which brings forth the orientation for the new culture, also means an advance in research. Life research, as described in Chapter 3, is an inherently significant part of the necessary expansion of consciousness. It has little to do with traditional science, for it constitutes research in a much more committed and existential sense and will operate with quite different methods. But it will retain and aid the idea of science. The idea of science is primarily the idea of objectivity or cognitive intersubjectivity. This idea is a true achievement in terms of humane progress (no matter how banal a level it has degenerated into in the scientific establishment). Belief, in the form it has taken so far, has been closed off against all appeals to reason and has caused too much harm on Earth for us to continue accepting it. Not belief, but knowledge – that is the true progressive solution of our times. And this does not mean only private knowledge, but public and societal knowledge. Thanks to its intersubjectivity, that is, its generally relevant, universal, and verifiable language, science is the *societal treasure* of knowledge that a people or a culture has access to at a given time. A collective substance is thereby created, which among other things unites individual history with the history of the species, and transforms humankind – after the breakdown of the old bonds of magic and emotion – into a social being. That intellect, knowledge, and cognition become a social force that can help people to regulate common problems outside of their momentary personal emotions – that, among other things is the human potential of science.

In the anti-intellectual attitude of current subcultures we see a major blindness to the evolution of the human mind, of history and of the patterns of development of the occidental human. Science is by far the latest and youngest bud that has opened on the tree of life, and we are not yet fully aware of its possibilities and riches. The intellect is ultimately – and this can be seen through precise self-examination – the authority with which we must justify our actions and decisions. Making this authority fully conscious and energised is of such paramount importance we can say unequivocally that, without science and without the intellect to use in its finest and most precise form, we cannot today attain cultural progress. The fact that cultural progress over the last two centuries was quite problematic – progress that lay mainly in the areas of intellect and science – was due not to the nature of the intellect, but rather to the nature of emotional structures that blocked the free development of the intellect in much the same way a free eroticism has been blocked. But no amount of misuse of the intellect in the sense of rationalising destructive behaviour, no amount of misuse of technology, and no amount of academic banalisation of thought will hinder the human from starting to think again, once he finds something worth thinking about. One reason why people in the present alternative movement often wholesale reject science and research is that they have rarely found positive areas in which to apply a new way of thinking. An important driving force in human existence – research and cognition – therefore remained unused. Not using an organ causes it to tire and whither away, as clearly stated by Hugo Kükelhaus. Not using our cognitive organs has, to a large extent, led to the mental and intellectual weariness and weakness we see today. But wherever new experiences turn life into a process of cognition, our senses open up again to the incredible aspects of our world. Every transformational community that seeks to prepare for coming epochal changes by developing a corresponding change in consciousness will, sooner or later, create a kind of research centre, for their mental-spiritual development. There they will study in depth the questions of a new kind of consciousness, a new mode of cognition and a new culture in a more systematic way. These research centres will be central points in the mental-spiritual network of the emerging culture. They, and not society's existing educational centres, will develop the qualifications necessary to address the questions of a new ecological society. For they will create the new human and social framework to formulate the right questions for the future.

EVOLUTION AND GROWING FREEDOM

The idea of evolution is that the totality of life on Earth is in a constant process of becoming. From the first cell to the human being of today there is a continuous line of development. The history of the human is the youngest episode in the history of evolutionary life and is itself a part of natural history.

The concept of freedom in evolution is the idea that the combinations that life has chosen in the process of its evolving did not simply follow a mechanistic determination or the laws of chance. Rather there was at the core of the living being something that could use chance, picking and choosing among the results of chance. For this we must assume a rudimentary proto-consciousness that from the beginning has directed life from within. This proto-consciousness, acting as the "impetus of the world" (Teilhard de Chardin), picked its way from possibility to possibility, from life form to life form, from species to species. This means that living matter (and probably all matter, considering the continuity involved) has a hidden centre that in the course of evolution is expanding in the direction of growing spontaneity and autonomy up to the freedom of will that we find in the human. It may be that the essence of evolution is nothing more than the development of this hidden centre now slowly being unveiled in the human. Essentially then, evolution is an attempt to realise ever more freedom within the medium of the material world. (Teilhard de Chardin has brought together impressive facts and thoughts on this evolutionary perspective in his book *Man and Cosmos*.)

This is not the place to expound on the metaphysics of evolution or to try to answer questions we are only just beginning to formulate. We are *living* – and this fact suffices for us to recognise directly the two qualities of development and freedom. Since we cannot believe that the miracle "Man" at a random moment in time was suddenly planted on Earth, we assume that these two qualities belong to the history of life and are inherently prepared within it. We assume that today's human is also a preparation, a phase in a total process whose goal and end we do not know. From this attitude stems our reluctance to make a definitive statement concerning the development of modern humankind with all its blind alleys and delusions. It may be that all this insanity was necessary to prepare for a higher form of consciousness. How can we know what the universe has in store for us!

Our thoughts about evolution do not simply constitute a private philosophical pleasure. They have decisive consequences for the way we imagine a desirable future. The fact that the ideas of a "New Age" often go hand in hand with the image of an egalitarian farming society and a simple nature-oriented life, free of technology, only shows that the eyes of most seekers today are directed not towards the future but towards the past. The evolution of life, which has achieved its most contradictory and enigmatic manifestation in the human, does not allow for a simple move back to nature or flight away from the tasks that our cultural era is currently failing to come to grips with. Instead we need to recognise the increasingly discernible primary direction that evolution takes – in terms of growing complexity, consciousness, and freedom – to rebuild and cultivate the Earth, utilising to the utmost our historical experience, scientific knowledge, and technical abilities.

The human being is at the forefront of consciousness as life evolves. He has no choice but to take increasingly conscious control of the direction that evolution is taking. This requires us to know or get to know the basic processes, tendencies, and operating principles of the living world and its evolution. This implies research, learning to see, that is, acquiring an attitude that enables us to see, in the sense used by Nietzsche and Teilhard de Chardin, and also by Kükelhaus. Wanting to lead a creative life today implies being aware of a goal. To set and realise goals with the help of our power of thinking does not originate from human hubris but is rather a human expression of the nature of will immanent in the Living. The freedom contained here has led to a constant shattering of energies. It is not until realistic goals, in evolutionary terms, are seen and understood that decisions can be made to join forces in rebuilding the Earth.

In the continuity that lies behind us through aeons of time, life developed from the formation of the Earth, and out of that life the human emerged. Geogenesis – biogenesis – anthropogenesis – these three developments constitute the direction of evolution to date. The world in which all this took place is described by Teilhard de Chardin as follows:

"In order to give birth to us it has, from its primeval beginnings, played in a miraculous way with so many improbabilities that there is not the slightest danger for us if we continue to put our trust in its guidance to the end of time. If it has taken on the task it will also be able to bring it to completion, using the same methods and the same infallibility as it has up till now."

Taking charge, and still trusting in something greater than ourselves – that is the logic of the behaviour that has emerged for us as a result of contemplating evolution.

A CULTURE WITHOUT SEXUAL REPRESSION

There are discoveries that normal mortals fail to make because what is to be discovered has become so strong a habit that it is no longer noticed. Freud's discovery of repression was of that kind. The importance of that discovery – as Reich's work testifies – reaches far beyond the conventional psychological area. Repression is the basis of our culture.

In adults repression occurs as fast as lightning and with the precision of a fully automatic instinct. It is therefore rarely possible to be aware of what is being repressed. Only through practised self-observation can one recognise what – often only for a fraction of a second – shoots through the organism. It may consist of highly energised and wonderfully horny images, fantasies, and impulses that immediately collide with the principles of so-called human dignity and therefore are repressed. The images are repressed and the energies are blocked through an irrevocable counter-impulse activated in this fraction of a second. If we – in a positive experiment in self-awareness – try to achieve the opposite effect, by holding back not the original impulse but rather by holding back the repression, then leave the organism alone, entirely free to develop its own pleasurable condition (e.g. masochistic or exhibitionist), its own excitement, its own flow of energies from within, and its corresponding world of images! Incredible sources of insight lie buried here! A whole series of amazing discoveries open up immediately. For example: a squeeze or a pinch that hitherto would have caused pain now creates a pleasurable sensation – how suddenly the body is free of pain, how obvious it is that sensitivity to pain is dependent on the mind! Or one feels fearless, strong, and at one with the self, in a state of deep and wholehearted affirmation. Life then responds from within with pleasure, confidence, and a sense of identity to the impulses that come from outside! Another example: one can experience a healthy feeling of omnipotence;

suddenly the body is incredibly supple, resilient, and as strong as a bear when the inner willpower no longer operates against the body but as its ally! What kind of communication and erotic contact would become possible here! The secret of the Living consists in its acting from within with a flowing force that would wash away all our troubles and sorrows if only we would let it. The secret of the cultural life of modern humankind consists in responding to this force with internalised automatic mechanism of repression. This was Reich's incredible discovery. The possibility of a fundamentally different culture is rooted in the truth that emerges here. Whoever has studied Reich's work or made similar discoveries may well face the most devastating fact that a perceiving consciousness can encounter. Imagine what we would know about humans if there were no repression! Think of all that we have concocted, puzzled over, and made up about the higher nature of humankind because of repression and because therefore the mind's eye could not see!

After the fall from paradise, God entrusted the cultivation of the Earth to humankind, but they did not accept it. They set forth to act as if the real and final solutions to human problems could be found in heaven and not on Earth. There was perhaps no time in human development when the human truly felt entirely responsible for all that he did on Earth. Herein may lie, in part, the great necessity and the positive function of the terrible wave that has been approaching us for the last 200 years and is now threatening to break: a fundamental change of command, whereby the human takes on all the responsibilities he once handed over to God and gods, spirits, the stars and fate. This includes the responsibility for what has seemed a matter of course in the traditions, customs and morals handed down to us. Only the thought that God is dead could bring the scope of this immense task into focus. This may be the decisive point towards which history had to turn. Perhaps the deepest thought in the history of ideas is that the human is responsible for himself.

All that has until now emerged and grown, more or less on its own – the form of the State, of the economy, of the family, and of social and moral order – should now be tested and cast into new forms by increasingly autonomous people. This also applies to the sexual order too, but the importance of this question could only be seen when Freud, through his concept of sexual "repression", gave us the keyword. One can only see what sexuality means if one has the presence of mind to recognise the inner impulses that are repressed practically at the moment they appear. Repression is the key process in the psychology of our culture.

But – as Wilhelm Reich discovered – it is also the key process of *physiology*. As a medical doctor, Reich could observe that the way the total vegetative and motor body systems function depends on whether sexual energies can flow freely or not. He demanded, in the name of health, that all forms of sexual suppression and repression be lifted entirely. This postulate was thorough and flawless. Whether we comply with it or not is a question not of taste or intellectual arbitrariness but of knowledge. So much pseudo-culture and hypocrisy, so many philosophical errors and political blind alleys have been caused by unexpressed sexuality that we no longer hesitate to say: *the possibility of creating a humane culture will exist only after sexual repression and sexual coercion have been eliminated.*

We say *possibility*, for it is in no way a guarantee. Here we are more cautious than Reich. He meant that removing all sexual suppression would also remove the causes for asocial behaviour, crime, perversion, and insanity. That salvo certainly overshot the mark, fired by the dynamite of a tremendous discovery that can hardly be handled by a single individual.

Reich's discovery is like other great discoveries and theories in that it must be intensively applied and tested over time before its importance and limitations can be seen. Today we can correct Reich's postulate and with the same intensity we can say: For a culture (even if it exists only in the community of an experimental village) to emerge without serious asocial behaviour, insanity, or crime, it is necessary, *among other things*, to free the sexual life of the participants from all prohibitions and slander. This freedom still offers no guarantee of humaneness, and is far from being its synonym, but it is a central prerequisite.

That we have become sceptical concerning an automatic humanising effect of sexual liberation is due partly to a growing familiarity with thoughts that were most poignantly formulated by Nietzsche in his "Genealogy of Morals". These ideas helped free us from the hypnotic and all too simplistic humane counter-images that a whole generation has carried ever since the students' movement of the 1960s. The human is much "worse" than we thought, and not only through the repression of urges and drives, but fundamentally, that is, based on his biological ancestry. The instincts towards power, plunder, ambush, destruction, revenge, and atrocity lie deep in his soul (although all these instincts are subject to evolutionary change and can change into something entirely different), and they have in no way been overcome in the evolutionary sense, at least not by the majority of people. The relinquishment of our hitherto morals and the freeing of sexuality that we consider indispensable will bring clearly to view the dark forces that lie buried in the deepest confines of the soul. Only then, when we see the hidden impulses of competition and intrigue, the gloating over others' misfortunes, and the struggle for power and love in their entirety can we begin to comprehend the scope of our question. The question is then:

How can we create a system of social conditions in which human energies can be channelled in such a way that they are neither lived out ferociously nor required to hide in a corsage of morals. In short: *How do we create humaneness that is not based on repression?*

THE MEANING OF SEXUALITY

Love is part of the magic of life, and sexual attraction is one of its most elemental forms of expression. In free – that is, uninhibited and unblocked – sexuality lies the elemental force and the depth, the passion, the calm, the resilience and the devotion of life itself. Sexuality is devotion, whether in an active or passive sense.

The fact that esoteric and religious teachings of liberation have usually been tied to sexual asceticism proves that sexuality has a special meaning in human life. Adam and Eve ate of the *forbidden* fruit from the tree of *knowledge* – and they "knew" each other (in Hebrew the same word is used for knowing someone and making love!) Therefore they had to leave paradise. What does this fantastic equalling of cognition, sexuality, and sin say about the essence of sexuality? It says at least this: that sexuality, and therefore the fact of humankind's polar duality and the ensuing striving for union, does not serve solely the cause of biological reproduction, but takes a special place in overall human development. The potential for insight and development inherent in sexuality is of the magnitude of early mystical knowledge, although we today choose to view it in a somewhat more dispassionate way. Sexuality is a fundamental issue in human destiny and human culture.

Whoever looks at or reads good pornography and does not brush it off through hasty indignation may find that it affects his body in a sudden rush. What are we dealing with here?

Whoever allows himself to be affected this way and gives himself over to the arousal notices that somehow it reaches his core. It must be an enormous force that can shake you in this manner. What is it? Whoever is so incautious as to expose himself to this power in the streets or parks or on the beaches; whoever senses that here something inescapable is demanding attention, and whoever despite an arising fear, or especially because of it, is determined to pursue it; whoever realises that here signals from a life unlived fire into our daily life; whoever no longer falls back on the old excuses and moral swindles in face of his longings; whoever is interested in and opens himself for this longing because he recognises in it the voice of an unfulfilled but real possibility in life; whoever is no longer satisfied by the explanation that all this is mere projection – knows something about the meaning of sexuality.

At heart everybody knows it. But we live in a secret complicity of pretences. For we live under conditions – involving marriage, reputation, social position, etc. – that would immediately be endangered if in the sexual area we came down on the side of truth. Sexual honesty would stand in incompatible opposition to our way of life, morals, science, religion, society, and also to our political and cultural customs. It would perhaps gradually face us with the most shocking of insights: that, considering our true wishes and longings, our daily lives are pretty ridiculous. For this reason the issue of sexuality needs to be forcefully taken out of the private sphere, if more people are not to perish for lack of fulfilment in their private lives. It is part of the perversion of human social history and the history of ideas that a phenomenon such as sexuality, whose bodily and emotional offshoots penetrate all of society like a secret nervous system, has been banished to the private sphere. Sexuality is a public issue of the first order and therefore should be treated so in any realistic cultural concept.

Sexuality controls and oppresses us as long as we do not succeed in bringing it into daylight and building a new social world with it. We can tie sexuality up in personal relationships, sublimate it in various activities, or turn it into mystical love – but some part of it remains outside, outside of communication, dark, strange, unredeemed. This residue that doesn't want to dissolve badgers us with a cascade of sexual stimuli which confronts us in every city in the form of a thousand breasts and thighs. The kind of excitement that grips us here (if we have not immunised ourselves well against it) clearly shows that in this central point we are not yet masters of our lives, not yet free citizens of the world. Real control of sexuality is a prerequisite for a true humanism and a truly free life. But real control of sexuality is identical with its total liberation, for only when it is free does it lose its pent-up power, its compulsion, and its underlying domination over our lives. Only total sexual liberation can free us from the secret tyranny of the unredeemed Satan.

Sensuality brings tenderness. We experience love when we hold a young rabbit in our hands, our hand transforming into a sense organ for living creatures. This sensual relationship to living creatures conveys the elementary feeling of love. Nobody would get the idea that there were anything inhuman in this process. Let us translate this to the relationship between human creatures. Under normal circumstances no woman who was touched in such a tender natural, and loving way would be shocked by it. Love would arise here at a creatural level, with the implicitness of a natural law if the flow were not blocked off through social rules, fear, a jealous spouse etc. Free sensuality and free sexuality need a new social order to be able to unfold their humanising force.

Beauty lies in the supple movements of an animal, in the erotic movements of a body that is free of fear, in the grace of a loved child and in the face of a loving woman. All ways in which life expresses itself are beautiful when the inner movement and outer expression are identical. Life is ugly and mean only when it is blocked, suppressed, and twisted. The human becomes ugly and mean when he has to suppress himself and pretend, because he cannot or may not commit himself to the life in him that pulls and hurts and for which he longs, whether he knows it or not. Sexuality becomes ugly and mean when lies enter into it. Blocked life becomes ugly and violent wherever it must force its way because its natural flow is blocked. Suppressed sexuality follows the principle of self-fulfilling prophecies. He who declares it to be a sin causes it to be a sin, that is, violent, asocial, and disturbing like a hidden ghost. To accuse pure sexuality of being animalistic is to accuse life itself. This attitude is caused by two enormous blind spots. First he fails to see the beauty of the animalistic. Secondly, pure sexuality among humans is never only animalistic, for the human is a mental-spiritual being. Guiding images of transcendence, of devotion, and of love are seething darkly in his cells until he has recognised and realised them.

The entire history of the patriarchal era is characterised by sadism, against which we cry out since we can no longer ignore it. The victims were raped, humiliated and slaughtered. Eros, striving for the union of the soul and the flesh, is an enormous driving force in history. Where the union is impossible the dammed-up energies turn into destructive energies. But even at the height of destruction the main urge shines through: in the midst of frenzy there is a transcendence of the boundaries, a dissolution of the perpetrator caused by the dissolution of the victim. We can empathise with it, the process carries a certain ingredient of lust and fascination. In the sadistic ecstasy both aspects come to bear: the aggressive blockage and the organism's longing for expansion, dissolution, and union. Both sides uniting in such a terrible way in sadistic excess, result from the division of the one and the same positive force that we call Eros. The fact that it is love, through human intervention, that brings forth its most terrible counterpart, that Eros reappears at the most extreme point of atrocity (in the form of sadistic fascination), is what moved Georges Bataille to speak of the "tears of Eros", and is what moves us to do everything possible to free our destiny and that of our children from this insane vicious circle.

The dissolution of boundaries contains an essential element of human change and transformation. Since the unredeemed Eros cannot accomplish this transformation in the medium of love, it does it in the medium of violence, either through open violence or in violent fantasies. But it must fulfil it, for the dissolution of boundaries, union, and transformation are its archetypal theme. At a deeper level all sexual forms, even the most perverse ones, have to do with this theme in its soul vibrations. If we understand this encoded theme of transformation, then we may assume that practically every form of sexual and sadistic excess existing today has already, somewhere, and at some time in human history, emerged as a mythological or cult element. The Aztecs had a ritual in which the priest removed the skin from a girl and wore a mask made of the skin of the victim's thighs. This was a fertility rite where the priest represented the son of the corn god and the girl represented the corn goddess. According to Erich Neumann in his book *The Great Mother,* the cult is an archetypal symbol of transformation and rebirth. The soul-like hinterland of sexuality is that deep.

The fate of the Earth is at stake. Whether we are to have war or peace is, among other things, dependent on the attitude of the human towards sexual life energies. What needs to be discovered is the simple fact that a free, fully affirmed, lived, and loved sexuality constitutes the catalyst of a humanisation of life. Love needs to be freed from "sin", that is, from repression, defamation, and lies, without losing the fascination it had as a sin. Life centres need to be established in which the opening of all libidinous channels can make it possible to build up an ecological culture from within. It is only in such centres that it will be possible to develop a mental-spiritual culture free from the compulsion to compensate for a life not lived, a religiosity free of hypocrisy and a love of one's neighbour, not based on morals and secret power, but on love.

A NEW SOCIAL ORGANISATION OF SEXUALITY

The social organisation of sexuality is a question at the core of every culture and society. But this question has seldom, if ever, been raised consciously in history. Its full meaning became visible only through the enlightening work done by psychoanalysis which, in particular, uncovered the subconscious effect of repressed sexual energies in all human relationships and social processes. That discovery produced – in a sense opposite to the view of Marxism – a fundamental new aspect in the understanding of human history and society. The decisive questions were no longer only in the area of economics or politics, but one level deeper in the areas of biology and sexuality. The so-called "societal basis" consists not only of the economic organisation of society but more comprehensively its sexual organisation, meaning the way in which a society organises its sexual life and channels its sexual energies (for example, every advertisement that operates with sexual symbols takes part in the channelling of sexual energies).

If we see sexual energy as a basic component of our life energies, we immediately recognise the importance of this question. The channels of sexual energies permeate the body of society like a secret nervous system that channels and transmits impulses of sexual attraction and repulsion. Whether this nervous system of society operates in the dark – below and outside social consciousness – or is an openly and consciously integrated part of social life is, of course, of utmost importance. In the first case it directs human life (including public life) from a region that lies outside social control and, therefore, is not accessible to conscious social influence. In addition, repressed sexual energies are almost always (owing to the repression) destructive in nature. Only rarely can they be sublimated into truly positive cultural feats. In the belly of every sexually repressed society an explosive mixture of sexuality and brutality tends to form and the monstrosities that occur as it ignites know no limits. Crusades, inquisitions, war, and concentration camps are among the psychological consequences of a culture falsely programmed in the sexual area.

The correlation between repressed sexuality and cruelty has characterised the whole course of history to date. Morals have served as a bastion against the asocial character of repressed energies. But this did not remove the repression; rather the repression was internalised. A vicious circle was thus set up that no religion and no peace ideology so far has been able to break. The humanisation of the world remains an impotent appeal as long as its biological prerequisites are not understood and created.

The social organisation of sexuality in our culture so far has been characterised by the system of marriage and family. One needs no psychological knowledge to realise that this system does not correspond to the natural needs of the human. But even the horniest faun, fully aware that he cannot satisfy his sexual appetite with a single partner, will swear by the institutions of marriage and family. There is a cultural mass suggestion at work here whose essential origin lies in the emotional situation of the child in the family. A child's experiences in its relations with its father or mother act like a post-hypnotic command to become just as they are, and unconsciously programs the conceptions of love and sexuality so strongly that they can hardly be altered by real needs or reason. The internalised family mythology projects onto every loved person a father, a mother, a daughter, etc. *The projection determines the choice of partner, leading to an unending repetition of the same structures.* Under these conditions there is no further creative development of love and sexuality. The evolution of humankind is to this day blocked in this central core of its existence. The mystical image of the one and only great love arising in the child through the mysterious glorification of the parents' sexuality directs the longings of the young ones towards fantasy instead of reality. *This fantasy remains the same from generation to generation.*

Variety is part of a free and creative life. A potent person does not remain at the level he has attained nor is he satisfied with one ability, one style, one method, or one theme. That goes for sexuality too. Every man and woman would understand that and would affirm it passionately, had our culture not deeply programmed us to see the renunciation of our drives as a moral obligation. It is necessary to keep in mind that we are confronted with immense suggestions stemming from an entire cultural tradition that has repeated itself for centuries in an almost unchanged form: the suggestions of the Old Testament, of Christianity, of the concepts of sin and the evil flesh. It is evident that it takes a special kind of work and effort to overcome this deep seated programming.

In the system of marriage and family, the fixation of the child on the parents and the Oedipal contortions of emotions, associating childhood love with fear and hatred, creates a character syndrome that, if we were to argue morally, would have to be characterised as "emotional mendacity or lying". But we want to discuss this question biologically. The biological (libidinous) energies of the child in the nuclear family system are twisted, turned against each other, paralysed, and blocked. The suppressed need for love, often frustrated through sexual morals or through the insensitivity or brutality of the parents, combines with the opposing energies of fear and hatred, leading to a paralysis that no longer permits free movement. The result is the source of practically all psychosomatic illnesses: a chronic emotional blockage.

The pent-up human – provided he has not found his specific pressure valve in his profession, sports or art – lives in constant energetic irritation, that is, he has a hard time concentrating on the things at hand, he is easily distracted and irritable. His unfulfilled wishes and drives force him, consciously or subconsciously, to a constant search for something else or to vicarious activities. No man can converse with an attractive woman about an interesting topic without being constantly distracted by her breasts. This is embarrassing but true. Our entertainment, recreation, and tourism industries make a living entirely from these concealed needs. In the sexual area we have the well-known games of extramarital sexuality that are programmed in part by the institution of marriage: hunter-gatherer sexuality, brothel sexuality and the endless search for partners. At the cost of an enormous amount of time and energy the trophies are acquired, trophies that satisfy no-one. If one considers the amount of fuel used only for driving around in search of a sexual partner, one would have to demand the liberation of sexuality solely for reasons concerning the politics of energy.

Parallel with more direct sexuality, a special kind of sexual underground has developed: pornography, peep-shows, personal ads, and special contact aids. (I speak myself without the slightest moral indignation or arrogance; I am myself always excited by such things.) If, in discussing the sexual underground, we include the entire area of latent sexuality that remains in the realm of fantasy, secret perversions, and silenced needs that are not acted out fully, then the accepted part of the psychosexual structure of society compares to underground sexuality as the visible part of an iceberg to the part that is submerged. The public system of marriage stands out in an amorphous, seething, restless mass of unfulfilled longings and withheld energies. Official life takes place on the surface, "essential" life lies beneath. *In this ambiguity of our culture lies its illness.* The necessary therapy therefore consists of a direct integration and socialisation of all repressed sexual and emotional energies.

Eros, being the biological and psychological principle of union and socialisation, is one of the most central, suppressed and, at the same time, most revolutionary productive forces of history. A new culture is rooted in new relations between the sexes, free of fear and projection. In a truly free and social human life there is no destructive fermentation, no remnants of mistrust and hatred, no waste of energies through pretences, and no need to project or vent pent-up images or longings, and no fear of the glance of others. *Such a free life will only be possible in a new sexual and social order where a sexual attraction between two persons no longer calls forth the fear of abandonment, paralysis or hatred in a third person.*

In art, philosophy, and religion, our culture has glorified love. But what happens to love as it is caught in the trap of a normal couple relationship? Almost with the certainty of a law of nature it turns into jealousy, blackmail, and boredom. The sexual taboo towards the outside creates inner dissatisfaction, aggression, and boredom. The partners have to hide this from one another otherwise their relationship and the comfort of their habits would be jeopardised. At heart, of course, they both notice that they are trying to keep up pretences that have long since ceased to be true. They become distrustful and suspicious. Whoever is unsure of his own love, also no longer trusts the love of the other. Now emotional war breaks out, a war that probably claims yearly more victims than traffic accidents. Behind jealousy there usually lies the psychological trauma of a child who fears losing the love of his parents. As the system of couple relationships is usually too limited to fulfil the individual's emotional and sexual needs, this fear of losing one another is constantly reinforced. If there is only one partner it is in fact truly traumatic to suffer this loss. A human system needs to be developed that allows for so many emotional and sexual relations and so much creative activity that the individual is no longer dependent on one person for the fulfilment of his wishes in life. Under such circumstances true love between partners could develop, of a freer and more beautiful kind.

Two tendencies in the historical developments of our time have led to the necessity of liberating sexuality from its old social forms: the process of sexual enlightenment initiated by psychoanalysis, and the inner process of destruction within the family, that was initiated by the development of the modern industrial society itself. The emotional erosion of personal relationships in the modern organisation of work, money, and consumption has long since affected the family. The daily "evening in front of the TV" documents it clearly: the family is no longer a source of creative liveliness but a refuge for passive recuperation and for working off stress, disappointments, aggression, etc. The emotional emptiness and confusion can be seen in the growing frequency of alcoholism, child abuse, and juvenile delinquency. Psychologists, sociologists, and criminologists are faced with a task that can no longer be solved with their limited professional methods.

A new order of emotional and sexual relationships emerges slowly. The removal of marriage and the nuclear family, while maintaining the other societal structures, would lead to total chaos. A common experimental living situation needs to be developed where, in free communication and in an atmosphere of growing trust, the forms of our daily lives can be *emotionally* overcome and replaced with new ones not based on ideology and outer norms but that stem from within. These new forms should not simply totally negate the old ones. The positive elements (love, warmth, trust, stability, clear orientation for the children) can be fully incorporated in new forms that are free of emotional constrictions and where the children no longer get caught in a web of fixations and projections. Of course such new forms cannot be planned programmatically, they must develop as a result of the relations between the persons involved. But within such a community smaller groups are bound to form, maybe in the form of smaller units or living groups of perhaps six to twelve persons. In these subgroups the "family" would then be embedded: the child with the father and mother as primary carers and the others as a care group. Children should have the possibility to choose their carers by themselves, because the *free choice of partners without fear and secrecy is the basis for emotional honesty.*

THE QUESTION OF NON-VIOLENCE

The dream of all humanitarian striving was always and remains the building of a non-violent society. But a person's attitude towards violence is not a question of his verbal commitment but of his emotional condition and the structure of his drives. Even Jurgen Bartsch (a paedophile and serial-murderer) would have been against violence if asked. The hidden tendency towards violence often lurks in the most gentle and quiet souls. It was well-behaved and unassuming family fathers who celebrated their bestial orgies in Auschwitz and other places.

In nature, in the evolution of life, and in the history of humankind, violence is such a prominent and multifaceted phenomenon. One might ask whether, in the entire process of nature, violence should not perhaps be accorded a more prominent position than our frightened souls like to admit. One might consider whether behind the desire for nonviolence there is maybe more personal fear than a grounded, realistic, well-thought-through vision of adult human relations. In any case, the problem of violence has not been thought through, because it immediately ends up in ideological systems directed largely by emotional forces that either serve to ease one's own fear or to defend one's own desire for violence. These two main aspects of the problem – violence as a basic element of the living world (Friedrich Nietzsche) and violence as a result of suppressed drives (Wilhelm Reich) – must be seen clearly and in depth before one can be allowed to take precedence over the other.

We have all watched a cat become fully awakened in a flash if something moves near by, and how it pounces to quash it. If we hold this process in front of our mind's eye long enough, then we make the surprising discovery that we *understand* it. Thanks to a deeply rooted possibility of identifying with the predatory animal we comprehend its preying instinct. If we allow it (if our inner censor frees this instinct in us), we make a discovery about humans through the observation of the animal, and this discovery is far-reaching. We understand, for example, that we could also hunt down and tear apart the rabbit that we hold so gently in our hands. During the greater part of history the human was such a passionate hunter and destroyer that it would be very odd if this characteristic were entirely eradicated from his instincts today.

When we attempt to create a new life-oriented culture we find that the biological and historical phenomenon of violence has some implications for a future world-view and ethics that would put it at variance with those commonly accepted today.

First of all, nature is beyond the categories of good and evil. A mysterious aspect of nature is that life is intimately connected with death, and lust often with annihilation. Secondly, in order to be able to fully see and accept this fact, different mental and emotional qualities are needed than those we commonly find in our culture. Especially needed is a fearless attitude towards pain and death. The ultimate solution to the question of violence can only be achieved from another cultural level where humankind has achieved a mode of existence that is fearlessly familiar with death.

Even if we cannot now fully clarify the question of violence, we can still commit ourselves to a decision. It is possible that Nietzsche was right in his theory about the inherent cruelty of life, to which would be added the cruelty resulting from human suppression of urges. This motivates us even more to develop a social system of living wherein violence, physical as well as that of the soul, can be transformed into constructive energies without being suppressed. We are appalled at the daily violence to which tribes, minorities, women, children, and animals are subjected. We are neither heroes nor cynics and therefore we want peace. In the centres of the new culture there needs to be peace, homeliness, warmth and security, and continuous development. If we are truly warmed from within through love of the living world, then our fellow humans and other fellow creatures will return this warmth. Then violence cannot so easily take on a life of its own.

Where the desire for peace has become serious and radical, we slowly start to understand that it can only be realised by fully overcoming all the life structures that instil latent violence by inducing the human to repress his urges and suppress his emotional energies. Wherever violence is due to blocked life energies, these energies must be liberated, including the energies of aggression. For here is the most important truth: *he who wants peace must stop suppressing aggressions.*

One can no longer disregard the emotional reality that so clearly emerges in every self-awareness group: the reality of our pent-up aggressions. Behind the façades of courtesy, tenderness, gentleness, shyness, and anxiety there is usually a huge amount of blocked and paralysed aggressive energy. We must face this latent aggression as a primary problem when we seek to humanise human beings and develop new communities. Latent aggression, developing from emotions and impulses never expressed, is a major cause of the listless life of boredom, exhaustion, impotence, and lack of communication to which millions have already hopelessly resigned themselves. It is of no consequence when a group of tired and resigned warriors proclaim non-violence on their banners and in their brochures; their emotional and energetic condition shows that they carry with them the true problem of violence entirely unresolved.

The "social character" of our time, that is, the basic character form common in most of our contemporaries today, consists of four different emotional layers that have evolved in the individual as a result of his inability to live out his urges and needs. The outermost and most recently developed layer is that of courtesy, good manners and unobtrusive behaviour. Behind that we find a second layer in the form of fear. Behind this layer there is usually an enormous (and biologically fully understandable and justifiable) rage. Only behind this rage, deeply hidden and usually well camouflaged, lies a layer of love and need for love that is quite childlike and vulnerable. It is in this psychological context that the problem of violence needs to be seen and solved. The tendency towards violence that is present in the rage does not, as a rule, appear on the surface, as it is paralysed through fear and camouflaged through conventional good manners.

This general emotional structure is of course closely linked to a moral-ideological structure and a social structure. Meaningless rage and violence are related, as fear is, to a state of constriction. They are a result of an inner and outer life situation that is too constricted and deprives the life spirit of the space and freedom it needs. The reasons for the aggression and violence inherent in our culture and society are, for example: the environment of a large city cast in concrete that gives the children little stimulus to activity; a career system where the division of labour creates the most meaningless monotony; a system of education devoid of real substance and without opportunities to follow true interests and curiosity; and a sexual system that keeps the energies of sexual love in the same old cages. These are the reasons for the inherent aggression and violence in our present culture and society. If we were ever to see how lovable and lively life is, how much charm and natural beauty, how much uprightness, courage, and honesty, how much trust and readiness to act are daily being betrayed, sold, and destroyed in the lives of children and youths, we would immediately dissolve into an ocean of tears. Even the resulting great biological rage has no chance to express itself in a meaningful way. People who grow up in this system must do so much that is meaningless and bad that they soon lose respect for themselves. This loss of self-respect then sets the perfect conditions for the ensuing insanities.

I am not exaggerating, but rather understating. If the word "peace" had not taken on such a piteous and hypocritical ring to it, I would say that the workshops of a new culture need to be workshops of peace. Peace will not come about until people have learnt to live in such a way that they can affirm themselves and others.

THE QUESTION OF DEMOCRACY

Democracy – are there any well-founded ideas about the meaning of this word and the possibility of its realisation? Do those who speak of "democracy" really want democracy or do they actually want something quite different? Many of the people who became the followers of a guru came from the political left and the alternative movement where it was natural to speak of democracy. The two movements that have caused an uproar in Germany during the last few years – AAO (an Austrian therapy and culture centre with free sexuality) and Poona (an ashram in India founded by the controversial spiritual teacher, Bhagwan Shree Rajneesh, or Osho as he was later known) – have at their centre an eminently charismatic leader. If one views the whole movement towards alternative living of the last 200 years, including the large American communes, one can draw a thought-provoking conclusion: so far, the only communities that have persisted for a long period of time have been of the charismatic type and not the democratic ones.

The political ideal fails, as always, in the face of emotional reality. The relationship between parent and child has nowhere been overcome. The grown-ups have not grown up. People have become sensitised to social and political conditions of oppression, they defend themselves against domination and authoritarian structures and therefore believe that they want democracy. But in truth (in the reality of their soul processes) most people defend themselves against authority for the same reasons that make them glorify their leader when they have found the right one: *they are emotionally fixated on authority, in fear and hatred as well as in love.* One should consider how much childish longing for love, how much inclination to adoration and worship could never be expressed in childhood because the adults were of no use – and what an enormous amount of latent longing is still there, waiting to be called upon! And then we meet a character who suddenly gives a green light to all these longings, one who symbolises the great positive mother or father figure onto whom the hidden inner child can project all *positive* attributes. Through this process true experiences of revival and awakening take place. The strongest force in the human being, love, is awakened. The libidinous upheaval is so powerful that it truly changes one's life. One has to have experienced it oneself – for example the way five thousand sannyasins celebrated Bhagwan's birthday in the Buddha Hall in Poona. Yes, it was like an enormous kindergarten, but it was a feast of flowing love, devotion, and thankfulness, in a way that the ordinary Westerner cannot imagine. When faced with volcanic eruptions of such genuine and intimate emotional force and identity (!) all intellectual nagging must cease. That is life. That is, at least for the moment, the truth of the common emotional structures that shape the true longing of most humans!

The catchwords for a better world – democracy, equality, non-violence, and social justice – almost always stem from crisis and resentment, not from a thought-through, positive conviction. That is the deeper reason for their failure. They do not touch the real emotional structures, problems, and longings of the people concerned. Wilhelm Reich drew attention to this dilemma fifty years ago in his book *What is Class Consciousness?* Much of what he wrote is just as true today.

As with non-violence, democracy is a question neither of verbal commitment nor of the outer political form of a system. Rather it is a question mainly of the emotional state and structure of drives in the human. Unfulfilled libidinous needs still stand fundamentally in the way of a free and democratic society. The emotional structure of today's human is not democratic and autonomous, but rather feudalistic. Just as in old times he longs for Father, God, and Caesar; but he does not want them in the old form, he wants a psychological equivalent for them. As long as no deliverer is in sight he does not recognise his inclinations and speaks of anything, such as democracy or even anarchy. But as soon as such a god-like father-figure becomes visible, he starts to come alive and forgets everything he preached the day before. I have often seen critical intellectuals, Marxists, ponderers and individualists arrive at the extremely hierarchically organised Friedrichshof in Austria and in very little time cease their resistance – not because they were broken down through brainwashing, as a sensationalist newspaper depicted it – but because they could no longer believe in their own resistance. Their true desires had been awakened. Here I remind the reader of the example of the two wanderers who went thirstily through the desert ...

Some readers may now understand me when I state quite simply that our culture of today, including our counterculture, is a pseudo culture. At the verbal level hardly anyone is credible any more, for people want something different from what they say. People are thirsty, but hardly anyone dares to say what for. The communities of AAO and Poona have brought this thirst out into the open, and what they teach us should be taken seriously. The political slogans of democracy, peace, and justice sound like Salvation Army hymns when compared to real life, as long as their psychological roots are not reflected upon deeply, down to the dynamics of their underlying drives and their emotional core, *and realised from there.*

True humanism needs democracy. All guru structures, all adoration of a leader, and all forms of organisation of human communities that are based on emotional fixation may be an important temporary learning phase for those involved, but they do not answer the question that we are faced with. That question is: what organisational form and inner constitution can we create for living together that can be applied generally and, in the long term, make humane structures possible?

Truly responsible humaneness can come into existence only after the fixations are overcome and the time has arrived when democracy is psychologically possible. The development of real democracy will be based on the reality at hand, for example, on the fact that in every community there is a kind of natural hierarchy (which can always change itself). Before the community can give itself a conscious form of organisation, some sort of group structure will already have evolved through the hierarchy of perceived human differences. These differences are a part of the variety of human biotope. They must not be suppressed through an overlay of egalitarian claims but rather should be used for creative learning processes.

Grass-roots and group democracy that reflects the Living is not based on egalitarian structures but on the optimum possibilities for individual development and growth in the intellectual autonomy of all members. These are high terms. They require the realisation of three things in the democratic society of the future:

Firstly, the child-parent fixation – which so far has held people in lifelong childish dependency on authorities – must be overcome through new social forms of raising children and new social forms of love.

Secondly, all emotional repression - which so far has stopped the emotional development of the human at an early stage and thus prevented him from growing up - must cease through a social system of free love, free research, and free work.

Thirdly, the greatest possible social transparency needs to be created (allowing the individual, from childhood on, to have an overview of his social environment, to know his present position in the community, and take part in current decisions). The next section considers some principles for achieving this social transparency.

Democracy cannot be achieved by fiat. It can only emerge and grow when the necessary emotional, mental, and social conditions are there. As it grows gradually and slowly, the community in which it develops will take the form of a circle. A circle where each element carries a different weight and is of different meaning but has its place and its relation to the whole.

BUILDING A HUMANELY FUNCTIONING COMMUNITY

Lasting and continuous cultural work requires building strong, supportive, and humanely functioning communities. Any attempt that does not consider this question is not a serious attempt. Too many political and alternative projects have broken down due to inner difficulties for this point to be treated casually. A solid community does not just spring into existence. Every group harbours the human problems, the hidden or open conflicts of authority, competition for recognition, power, and love, the seething aggressions, and the entire swamp of unsolved sexuality, jealousy, fears, etc. In all groups you find the whole spectrum of typical problems. The men often compensate for their emotional and sexual problems through intellectual rhetoric and ideological self-importance. The women don't really trust each other whenever men are involved. Those in love sexually often do not know how openly they dare show their wishes without causing fear, jealousy, and chaos in the group. In couple relationships, what was once love gets mired in a distrustful clinging to each other. Children never quite know whether they can trust their parents' love and get confused and absorbed in testing them constantly. In order for these daily occurrences not to perpetually repeat themselves we need new experiences and new knowledge about the inner composition of a community. In simpler cultures communities emerged naturally on their own. Today they need to be initiated consciously according to all the knowledge and experience of interpersonal relations.

Building a humane community usually means confronting difficulties that are deeply rooted in the character structures of modern man and especially in the ideological structures of the subcultures. Instead of the fixation on humanitarian slogans and demands, what is needed is a clear attitude towards the emotional and mental realities that in fact exist. This emotional reality must at all times and places be made visible, with as playful and joyful methods as possible, until all pretence and hypocrisy drop away. The group members must notice that there is no longer anything to be gained by pretending. There is perhaps only one categorical imperative for the emergence of communities that are good and stable from within: that everything possible be done to make what happens in the community understandable and transparent to everyone. This is especially important for the emotional and sexual processes, for they are behind almost everything that makes the group situation difficult and opaque. The transparency of all processes is the precondition for freeing the members of their paranoia, for keeping destructive processes from taking on a life of their own, and for treating the causes of rifts and fractures in the group before it is too late.

A crucial part of overall transparency is the transparency of social hierarchies, which exist in every group. It is good if everyone knows as precisely as possible his place in the group, what the others think about him, and where they see him in the hierarchy of the community. In this way the community is freed from the hypocrisy of superficial harmony and false democracy. Each individual can then get to know the reality he has to deal with. It is no longer easy for him to inhabit a world of secret claims and blackmail in which he used to be able to blame others for his weaknesses. Now he knows his place and can work from there.

Ongoing social feedback between the members of the community is also part of transparency. The members must learn to tell each other without hesitation what they like and don't like. Personal conflicts must not be suppressed or they will seethe under the surface, poison the atmosphere and lead to camouflaged reactions that cannot be read clearly. More serious personal conflicts should be presented publicly to the group. To avoid taking it all too seriously, those involved should learn to distance themselves from the problem through playful methods (such as psychodrama or the methods of self-expression developed in Friedrichshof). Conflict often turns out to be a part of one's own insanity, which in itself justifies perceiving and presenting it as a stage play rather than as a too serious reality. (To learn to play the games of group dynamics it is wise at the beginning to enlist experienced "neutral" therapists to teach the methods and something of the processes involved.)

A common cause for the lack of transparency in a group situation lies in the mixing of factual discussions with emotional conflicts. The group must learn to distinguish sharply between them to keep discussions from sinking into an aggressive quagmire. There is no point in continuing a factual discussion that has long since become dominated by personal conflicts, covert power struggles or some old story of competition. This is the time to break off the discussion and carry on the interaction through play-acting with theatrical exaggeration and playfulness.

Another central point in the building of a community concerns the question of couple relationships and of sexuality. Again and again the necessary transparency is severely impeded by the old structures of the couple relationship. As soon as difficulties arise in the group the partners tend to retreat to their coupledom, creating a private fortress of protection and resistance. The harm that is done not only affects the community but also the two partners. This is because couple relationships tends to be far too narrow to integrate and hold everything that moves and troubles the two partners. Jealousy especially cannot be worked out by them alone. An inner group coherence based on transparent structures can only grow and develop lastingly if the pair can dissolve the habitual boundaries and work on their internal difficulties in the group (as far as that makes sense). That such a dissolution of boundaries may also lead to new emotional and sexual contacts is natural and desirable. A community will only reach its human goal of solidarity and mutual trust if the members are willing and able to realise the old values of love – intimacy, trust, readiness to help, and partnership – not only with one partner but with as many as possible. In such a process of building community with increasing inner coherence, maintaining emotional and sexual barriers becomes increasingly senseless.

For several years we have carried out this experiment in the *Bauhütte* project (see Appendix) and we have found that mutual distrust slowly disappears from the relationships when each partner openly sees and knows what the other is doing. Openness and transparency in the emotional or sexual areas are crucial elements in the psychological structure of a new culture when it comes to clarifying the group situation. Partnership and free sexuality are not in opposition to each other, rather they complement each other and need each other. Free sexuality, based on lasting personal relationships, is no ideology or program, but a natural way to deal with reality. Admittedly it is just somewhat unusual after several thousand years of monogamy. But we are not seeking to maintain habits, we want to solve problems.

EMOTIONAL CLEANSING AND DISSOLVING THE CHARACTER ARMOUR

Earlier in this book a few basic characteristics of the new culture were described, without which a new order of humane society, oriented upon the laws of life, can hardly be possible.

These characteristics are: solidarity with all living beings, life research, free sexuality, non-violence without repression of aggression, grass roots democracy through transparency (especially in the emotional and sexual areas), and constant social feedback through direct human contact.

These characteristics cannot be realised at the depth of their true meaning on the basis of present-day emotional, character, and ideological structures; in part, they cannot even be understood. They seem to go against empirically determined laws of human behaviour. But in reality they only go against the empirical laws of the existing cultural era which is based on bioenergetic self-suppression. The realisation of the above-mentioned characteristics requires an inner process of change for the individual we could term "emotional cleansing".

What does "emotional cleansing" mean? It means that love and hatred are freed from their mutual embrace; that one feels no fear when one needs to fight, and no inhibitions when devotion is called for. It means that one does not force a smile when one would rather cry or scream, that one learns to differentiate between love and the need for someone to lean on, between a "yes" that stems from the heart and a "yes" that stems from the fear of being rejected. It means that one no longer confuses one's lover with one's mother or father; that one does not confuse emotional hypersensitivity with love of one's neighbour, or the rage of being personally hurt with the rage against the destroyers of life, and one's own cowardice with consideration and tolerance. Emotional cleansing means that the emotions and energies can flow again because they are free from hypocrisy, that the feelings of inferiority and guilt disappear because the inferiority and guilt have disappeared. It means getting rid of false feelings of shame with which we have denied our best and most vital urges, and that true shame emerges, the shame over our constant repression of the truth of the living world within ourselves and others, for no other reason than for our fear of the eyes and judgement of others. Fundamentally, emotional cleansing means to overcome the entire psychological and cultural system that Wilhelm Reich called the character armour.

The character armour is both a system that keeps down the biological energies and a psychological-ideological protection against all invasions of life that have been forced out, and against all signals from buried truths, longings, and love. The cultural era of the character armour has declared all grapes that hang too high to be sour and hated everything sweet that was unreachable. It despised and rejected the lust for which it had always longed, made impotence into the virtue of abstinence and turned cowardice into morals. This mendacity has become a solid structure and a permanent part of all that has been handed down as "education", "humaneness", and "human dignity". People instructed others about freedom and did not see the trap in which they themselves were caught. They developed theories as an excuse for their own fears, attacked the state and society but resisted every attack on their own character armour.

For the regulation of their social lives people of the old culture needed external ideologies and authorities. By being armoured beings, they could not rely on the honesty and reason in the human feedback they got through contact with their peers. Since they also were pent-up and full of latent cruelty they had to protect themselves against asocial excess through a system of laws and punishment. Fear has therefore been a central element in regulating society. If there can be said to exist one single central change of paradigm for our total culture, then it will be anchored in the *change-over from a social order regulated by fear to a social self-organisation rooted in free and direct human contact.*

Of course this change cannot occur overnight, not even in small model communities. But centres and support groups need to be established which, through their social and psychological structures, will be able to facilitate this central process of transforming human society. Using all available human and sociological intelligence we need to replace fear as a regulating principle by something we could call … love. The "home", of which the philosopher Ernst Bloch wrote, actually does lie in love, in a free, unsentimental love containing no remnants of fear, lies, or hatred; and the great *nondum*, the unredeemed part of history, lies for now in this, the greatest of unredeemed human longings. When it is fulfilled, when the human is loving in full sensuality, stands awake and fully present in the world, then finally a mode of existence will have been realised that we have always at heart known exists.

5 POSTSCRIPT

CONCERNING TRADITION

We are attempting something radical. It is aimed at what may be the deepest possible reversal of the values of the historical past. Once again human life would begin a new history. But the careful reader will realise how much history, how much conscious integration of tradition, and how much "conservative" will is present in our radical attempt. Since this is a radical cultural concept, that is, one that goes to the roots, we touch upon the most distinguished teachings of the past, even where we are diametrically opposed to them. Here, as everywhere, the "paradox" of life is confirmed. It may seem like a paradox to some that a project so close to the ideas of Wilhelm Reich can simultaneously quote with joy the words of the Jesuit father Teilhard de Chardin. This original mind, whose radicalism proved itself as much in his personal as in his scientific approach to the philosophy of evolution, calls the unspeakable apostle Paul a holy man. Paul and Augustine together are the philosophical source of the appalling form of Christianity that sees its mission as the fight of spirit against the desires of the flesh and in making the authority of the church absolute. Thus they both contributed decisively to the human catastrophe of the Christian West. They belong to that part of humanity who have decisively and deeply influenced the value judgements of our entire culture. We are formulating the most radical counter-position possible today – after Marx, Freud, and Reich. We must take care to not simply negate the old; it is much too *true* for that. Augustine was no weakling, nor was he a fanatic who spoke out of personal need or resentment. He possessed what the material-sensual world had to offer and yet he gave it up. Why? Because he was moved by a truth that is as deep as ours: the truth of "God", the truth of a religious love that is entirely freed from and cleansed of the flesh. This truth is enormous. If it were not, the belief that the body is a prison for the soul could never have held sway in the history of religion. Plato's eudaemonism (highest bliss), which also calls for separation from the body, and the flagellantism of the Middle Ages have the same origin, and that is the deep human experience that the body is a prison. This experience is spiritual in nature and therefore, in principle, lies outside the area of classical psychological explanations and theories. A truly relevant spiritual approach would find that the body is in fact a prison for the soul. As Rudolf Steiner said, many illnesses occur because "the soul is too closely bound to the body". The materialistic consciousness that reduces mental and emotional processes to bodily ones therefore causes a disposition towards illness that in the end can only be overcome through a new medicine and life practice that is religiously oriented. A cultural renewal based only on what is sensual and vital would once again miss the essence unless the functioning of the sensual and vital aspects of life were understood fully. They can only burst into bloom when we reach a certain level of "de-materialising", spiritualising, and releasing of the body from its inertia. The body will only become a full medium for sensuality and self-realisation when the mind is no longer fixated exclusively on this body and its needs. Mastery of the mind over the body remains valid though seen in a different light. Mastery no longer means suppression and mortification.

When today we dare to seek an existential orientation and to live according to it, we do not draw a line between thinkers who are close to our philosophy (verbally), and those who are far from it. The ideological battles that result from such superficial comparisons are meaningless when it comes to making basic decisions of a personal nature. The determining factor is the "break-in of the existential" into people's lives and the power of their ensuing inner revolutions. Those were the "originally shaken" (Karl Jaspers) – people such as Augustine, François Villon, Kierkegaard, Nietzsche, Teilhard de Chardin, Wilhem Reich and Otto Mühl have something to say to us, even if their truths must first be uncovered and seen outside the context of their specific historical form. Their lives and the nature of their work are virtually proof enough for us to take seriously that which moved them. Paul would not have been able to race through half the world if his idea of salvation had only grown out of his "personal issues" (such as the projections of an urge-adverse and pent-up soul).

The truth we are dealing with is never a finished product; it remains in constant historical development. Its aspects are as multifaceted and contradictory as the twists and turns in the history of ideas. Next to Nietzsche's philosophy of the will to gain power stands, equally valid and necessary, the Sermon on the Mount; next to the adoration of sensual beauty we have as equally relevant the ideal of chastity and asceticism, next to Marx's materialism, we have, of equal importance, Plato's "Idea"; next to Reich's sexual economy there is Buddha's concept of Nirvana; and alongside the modern idea of self-realisation, we have the equally profound Augustinian or Jesuit ideas of obedience. Again I repeat: we must meet these ideas – which in their verbalised forms often appear contradictory – in their essence, in what is truly meant, in order to be able to assimilate them. Take for example the religious idea of obedience to God and Church. Its essence is in the becoming capable of devotion, of giving oneself completely. For the human to be able to receive the holy spirit and thereby partake of love, a disposition must be created in him that makes him open to this receiving; and that takes the form of unquestioned obedience. (That the call for absolute obedience towards the Church was also motivated by reasons of power, as described for example in Dostoevsky's *The Grand Inquisitor,* is not denied at all. Whenever the authority of those who had the original experience was no longer effective, the process of perversion started.)

The ordained forms of asceticism, chastity, and obedience can only be convincingly surmounted and replaced by more humane forms when we understand their existential meaning and their spiritual depth. The same is true for the forms of marriage and family. They were all historically productive at one time, in that they addressed and developed an essential aspect of the human soul. We can only truly overcome them by understanding and preserving their true nature (this was the meaning of Hegel's term "sublation").

We have found the functional principle of opposites to be a basic characteristic of the living world. This "paradox phenomenon" sharpens our senses for the contradictions in history and in our own development. The structure and dynamics of our own situation is full of contradictoriness; because the impulses that life sends out in the form of our drives, needs, and consciousness are contradictory. Therefore we need to guard ourselves against hasty judgements and the one-sided choices of our philosophical reference points. The mental-spiritual physiognomy of what is deep and genuine is not bound to just one time, one generation, one ideology, one movement, or one assertion. Whatever is most profoundly contrary to our convictions may sometimes lie very close to us – for example Plato's theory of ideas, removed as it is from any life practice; the life-negating mysticism of the East; the idea of asceticism; the concepts of sin, grace, and forgiveness; the idea of marital faithfulness and devotion unto death; the idea of making sacrifices and giving up the self. For the living dialectic of opposites creates inner points of contact in places where the superficial eye sees only contradictions.

We formulate our radical thesis not outside our cultural era, but consciously within the deeper lines of development that we recognise behind the obscurity and perversion of our cultural tradition. The truths that are hidden in the old forms of religion, philosophy, science, art, love and social order shall be freed from these forms so that we may become capable of re-experiencing and developing them at a new level of consciousness.

Life has encoded its truths on the great stage of history and the history of ideas. Its cast of characters include Dionysus and Jesus Christ, Napoleon and Elsa Brandstrom, Francis of Assisi and Nietzsche, Rudolf Steiner and Wilhelm Reich (that hardly any women are included is due partly to the peculiarities of male historiography). The originality and authority *(auctoritas)* of the living world incarnated here can be seen in these people's actions, in the passion and authenticity in their work and style. Style, seen as the statement of something from within, is not a question of technique but of originality and proximity to the issue. Anyone unfamiliar with Nietzsche or Teilhard de Chardin would be taken aback by their styles – if he is sensitive to them. Whenever a distinct and great style emerged in the epochs of art history – the Antique, Romantic, Gothic, Renaissance, etc. – we can assume that it coincided with an inner attitude towards life relating to an essential mode of experiencing life. This of course applies to stringent moral codes. Morals have not *only* imprisoned the human animal in a self-imposed corset but also and perhaps mainly are an historical impulse towards self-assessment and self-education. In that sense they are a true humanising medium. That moral codes as a rule achieved the opposite is an indication not of the poverty of their motivation but of their inadequacy. Are not our own endeavours somehow of a deeply moral nature?

History has been a battle between the principle of love and the principle of fear. The original and authentic cultural creations have always been the attempt of humans to assert themselves in this battle. In figures with the highest level of consciousness in the history of morals and religion, the battle was fought for the principle of love. The modern form that this battle has taken – expressed most clearly by Wilhelm Reich – is based on a more precise form of analysis and self-observation, rather than on seeking a final solution. By using Reich's results we hope to avoid past mistakes in our current attempt, but they cannot protect us from making new ones. But perhaps, beyond the search that is unrewarded and stumbles into new mistakes, a new discovery can suddenly be made. Philosophical thoughts about liberation are powerless if they do not promote development in which an inner leap can take place that leads to lasting new experiences. We can construct a suitable vessel, but we can only hope that it will be filled. We want to build the most realistic basis we possibly can for the "principle of hope". Ernst Bloch has vividly demonstrated in his voluminous work that there is incredible material from the history of ideas that we may understand, transform, and assimilate. This way of studying history is for us a never-ending source of discovery and self-recognition for it is identical with our own process of becoming. The "psychic structure" that we have deep within us is – when seen as the precipitation of human experience – sedimented history. Self-affirmation at this level of consciousness is therefore also the acceptance of the tradition from which we come.

THAT ALL THIS DOES NOT REMAIN MERE WORDS ...

One reason for publishing this book is our hope of expanding a cultural project that has been in preparation in Germany since 1978. The project bears the temporary name *"Bauhütte* – Workshop for Life Research and New Culture". It was started by some men and women who for this purpose gave up their previous professions and put their savings into the realisation of the project. The goal is to build a exemplary settlement where the cultural concept outlined in this book is realised as a model. In order for a community of 100 to 200 people to function there must be not only a concept for the psychological and social aspects of living together, but also functioning concepts in energy, food production, and health. Also needed is an architectural concept that corresponds to the social concept, a sensible solution to the problems of solid waste and waste water, a system of pedagogy and education that reflects the contents of the new culture, and a system of its own for research and study. As much as possible model solutions in all these areas are being sought for the settlement, solutions that may come to be important examples for a future ecological culture and way of living. The new culture grows out of positive examples, and the existence of an ecological settlement with a functioning community and functioning technologies would definitely be such an example.

The co-workers of the *Bauhütte* project have developed a differentiated concept for the areas of energy, agriculture, and waste recycling that attempts to combine the practical needs of the community with the intention of creating a model. Questions have arisen that cannot be answered now, nor in the foreseeable future. Technical experiments are being conducted that will bring more insight. For anyone who gets joy from such practical research there is a wealth of research fields open for creative pioneering work. Many of these fields have lain fallow for lack of individuals who could cultivate them. The fields being actively pursued are: energy research, laboratory research for the study of biological energies and biological microstructures, research in recycling with an integrated combination of elements (a functioning plant has been built for producing biogas and algae, integrated with aquaculture and fish raising), air and water purification techniques developed by the physicist Ronge, and initial experiments are in process for building with new forms and materials. In addition, a special kind of "therapy research" has been carried out over the last few years to facilitate the dissolution of the character and body armouring. With this type of research at its base, a small medical centre will be established, for which we are urgently seeking dedicated doctors. These are some keywords for interested people. Those who want to know more details may refer to the brochure describing the entire project. Some *Bauhütte* co-workers are working on a very extensive basic research project that we call "life research". It is divided into philosophical research (addressing such concepts as biological "self-organisation", "information", "entelechy", etc.), scientific research (for the study of biological structures, forms of motion, functional principles, biological principles of construction, biological energies, processes of germination and growth, principles of ecological systems, etc.) and finally a type of human research to become clearer about the meaning, goal, and possibilities of our existence.

Thus the planned settlement is to be neither a therapeutic centre nor an ecological village in the customary rustic sense, but foremost a centre for new insights, experiences, life research, cultural research, community research, environmental research, and experimental social design. The final criterion for everything, including the new technologies or the questions of economic autonomy, remains the human dynamics, the ability of the community to love, and the individual's possibility for development. The cultural and political character of the project means that the practical concepts that are developed go beyond the limits of usual village concepts. Taking an active role when it comes to the questions of our time is part of a creative life. Increasing publicity is resulting in a network of political, cultural, and personal contacts that facilitate the realisation of the project and that may, in the future, allow the solutions developed in the community to have some influence on society at large.

Our project is not intended to fulfil a pre-ordained program, but rather to initiate a development from a solid basis. The project will grow to the extent that contacts and relations grow among the people involved. Our chance for a political effect lies solely in achieving human understanding and resonance. For the implementation of the settlement, we neither need professionals nor academic titles (although we also have nothing against them). We need people, co-workers and friends, whose joy it is to risk something unusual while using their intelligence. The more of them the better.

6 APPENDIX

TAMERA MANIFESTO
For a New Generation on Planet Earth
By Dieter Duhm, Tamera, Portugal, 2011

A quote from the Gaza Youth's Manifesto for Change from December 2010:
"We are scared. Here in Gaza we are scared of being incarcerated, interrogated, hit, tortured, bombed, killed. ... We are youth with heavy hearts. We carry in ourselves a heaviness so immense that it makes it difficult to us to enjoy the sunset. ... There is a revolution growing inside of us, an immense dissatisfaction and frustration that will destroy us unless we find a way of canalizing this energy into something that can challenge the status quo and give us some kind of hope."

This is the cry of the youth of Gaza. It is the cry for help of a generation without hope. A call from many countries of the earth. Representatives of Tamera Peace Research Centre in Portugal have undertaken several pilgrimages in Israel/Palestine and Colombia. May the following text contribute to a way out of the misery, and to finding a new channel for suppressed energies.

We greet the youth of the world. We greet all peace-activists and helpers in the crisis areas of the earth. We greet those who, often risking death, dedicate their lives to uphold human rights, for the protection of children and indigenous peoples, the protection of animals, the protection of oceans, trees and all co-creatures of the great family of life. We also greet those governments who still have the courage to stand up against worldwide globalisation and its methods.

This is a manifesto for a young generation which no longer has a future in existing society, for those who are actively involved in the struggle for liberation, for the relatives of the victims, for the unbearably many people who can no longer see a way out and who have no perspective in the face of daily misery.

The world is in transition towards a new way to live on Earth. The old dictatorships and hierarchies cannot remain much longer. We are experiencing the collapse of the mega-systems. The revolution in the Arab countries, the youth rioting in the Western mega-cities, the world financial crisis and mass unemployment, the rise of wars and man-made natural catastrophes, the moral decline into squalor of most governments, the international plans for states of emergency and the underground bunkers for the wealthy are sure signs of the approaching end of a violent epoch. Behind the global violence, powers of a profound change of era are showing themselves. Those who stand against despotism today could witness a completely changed world tomorrow. We greet those who are preparing the new era on all continents today, often risking their lives. We greet the newly arising planetary community.

GLOBAL SYSTEM-CHANGE

Behind the global massacre of our times stand wrong systems of economy, wrong concepts of love and religion, wrong systems of thought, and the endless abuse of natural resources. A global matrix of fear and violence has developed because of a wrong turn of evolution, and has eaten its way
deep into the collective human soul. The new planetary community is making a fundamental system-change from the matrix of fear to the matrix of trust. It is doing so in all areas – from personal relationship issues to the political and ecological issues of the healing of the planet. Most natural catastro-

phes are the result of wrong human intervention in the cycles of nature. This system-change is a change of power. The new power no longer consists of domination over others but of reunification with the sacred laws of life. Everywhere that destruction is now raging, the first cells of a new world are emerging. The global apocalypse, horrible as it is, means not only downfall but also revelation. The forgotten sanctity of all life is rising from the ruins of the old and now giving birth to a new epoch. The new communities enter the service of life, service in the vineyard of God, and cooperation with those highest powers who have always, from the very beginning, formed our universe. The system-change will happen surprisingly quickly. In a few decades our children and children's children will know the millennia-long age of war only from history books.

The Earth can be healed. There is a world that heals our wounds. This is the world of undistorted life. And there is a world that causes the wounds: the world of the human being. These two worlds have to come together to prevent future suffering. The world of the human being has to be reintegrated into the basic structures of universal life. The following four basic areas must be healed: energy, water, nutrition and – love. These four sources of life must be liberated from the dark powers which have destroyed them (energy companies, dictatorships, churches and so on). This is not a private, and not a local fight. It is a global fight. It is a fight between the global powers of life and the global powers of destruction. If life wins, there will be no losers.

A NEW PLANETARY COMMUNITY

Besides the global riots, there is today a global movement to save life on earth. It consists of groups from indigenous and religious peace-traditions, especially in Latin America and Tibet. And of those moved peace-activists, environmentalists and seekers of life that has long known that there is no future worth living within the existing systems. We see a new generation of pilgrims from all countries travelling across the earth. They are no longer bound to nation, language, race, culture or religion, nor to wealth or possessions. They help in crisis areas, visit sacred sites, meet at campfires and in hostels, share their bread and develop a new quality of community.

A new global citizenship is developing beyond all institutions. A new form of positive 'globalisation'. This process is supported by the development of new centres which slowly spread across the earth. We call them 'healing biotopes' or 'peace-villages'. They serve for the pilgrims as shelter, study centre and workplace. Real research work is done here on the technological, ecological, social, spiritual and intellectual basics of a non-violent world society. These centres follow a common ethic of living together, a charter of human rights and animal rights, a kind of planetary set of 'precepts'. The following eight peace-thoughts are valid at all places on earth:

1. HUMAN RIGHTS AND ANIMAL RIGHTS

They recognise basic human rights independent of religious or racial origin. They do not tolerate the presence of hatred, violence or humiliation.

And they recognise basic animal rights: the right of all animals to living space, food, freedom of movement, curiosity and contact. Animals will not be mutilated in the new world. No dogs will have their tails cut off, and no animals will be subject to experiments by the pharmaceutical industry. There will be no fur farms and no slaughter houses. Animals are natural co-operation partners and friends of the human being in the great family of life. Animals need our support and not our persecution.

2. THREE ETHICAL PRINCIPLES

They follow the basic ethical principles of community, particularly the three principles of truth, mutual support and responsible participation in the whole. In communities which are based on truth and mutual support, a power develops which is stronger than any violence. It is the power of trust: trust between men and women, trust between adults and children, and trust between humans and animals. The re-establishment of an original trust, in a world in which fear no longer exists. Trust is the basis on which life heals. There is no deeper vision than the vision of a world in which trust reigns between all beings.

3. SEXUALITY, LOVE AND PARTNERSHIP

They also follow the principle of truth and mutual support in the areas of sexuality, love and partnership. There cannot be peace on earth as long as there is war in love. The new world has overcome all forms of the fight between the genders. Neither chauvinism nor feminism exist. The genders stand as equals beside each other and work together for the same goal, the reunification of life. Questions of monogamy or polygamy, of couple-love or free-love are not ideological or religious questions, but rather questions of personal development and the decision of those who are involved. Love is a natural process, not a legalistic issue. There is no legal claim on love, or right of ownership of a love-partner, but there is great trust and deep solidarity between the female and the male halves of humankind. Sexuality is liberated from all forms of religious suppression, lies, humiliation and violence. It serves – in addition to reproduction – only mutual love, health and joy of life. In a humane world it can never occur against the will of one partner.

4. NO RELIGIOUS BARRIERS

There are no religious barriers in the planetary community. The same God, the same heaven and the same cosmic order of the sacred matrix reign over all religions. The sacred authority we serve is not a clerical institution, but life itself, because this is what we love. The Divine no longer reveals itself in old bibles, but in the movements of a stream, or in the construction of a blade of grass, and particularly in love and the mysterious interplay of the powers out of which all life originates. The Creator is not a punishing father-God. It is the I-point of the world, in which all vibrations come together. This I-point exists in all beings. When we meet anew in this understanding, there can be no religious violence.

5. GRACE: NO REVENGE. RECONCILIATION.

The newly forming planetary community has given itself a name. Their posters say "GRACE – Movement for a Free Earth." They are declaring that the injuries and pain suffered will no longer be answered with hatred and violence. The pain has given birth to a new determination. Hatred has transformed into an absolute decision for life, peace and healing. There is no neutrality any longer, as one has taken a stand for life. This is no ideological or political stance. The tears which are cried by an Israeli mother for her murdered son are the same tears as those cried by a Palestinian mother. For many, the pain is too great for any more tears. Accusations and judgements are of no use any more because they perpetuate the downward spiral of violence. The young demonstrators in Cairo or Tripoli are the same age as the soldiers and police shooting at them. They could have been friends. The peaceworkers in Colombia's San José de Apartadó and the murderous paramilitary could also be friends if they could step out of the constraints of a vile system. No revenge! This was the appeal of a young Israeli woman (Michal) after her face was destroyed in a suicide bombing by a young Palestinian. She said that she might have acted similarly if she had been in his position. The inner power of this attitude is based on the deepest insight that we humans all come from the same source, have passed through similar suffering and are striving towards a similar goal of peace and healing.

6. LIFE WITHOUT FEAR

There is no more fear of any enemy, as there are no real enemies any more. The Indian philosopher Sri Aurobindo fought as a revolutionary for India's independence from England. While he was awaiting a death sentence in prison, Vasudeva (God) appeared to him. Vasudeva appeared to him in the form of the guards, the prosecutors and the judges.

He had no more fear and was released. This is a very high point in the development of consciousness. When a human being has reached an inner point where he no longer reacts with fear or hatred, his organism changes and he becomes increasingly immune to attack and invulnerable. There are astonishing examples of this miracle. The Jansenists in Paris in the 18th Century could not be killed, as they were no longer afraid. In the times of the Great Plague, those helpers who were not afraid of infection were immune to the disease.

There are stories of people in concentration camps who were spared by the executioner because they were not afraid of his power and cruelty.

There is an essential key for peacework in our times here. Those who do not project onto evil cannot be reached by evil. Evil does not have power in itself, but gains power from projections of fear. An evil regime cannot remain in power once people stop projecting on it. It depends in the deepest sense on ourselves whether we win or lose this fight. We will win it once we stop reacting in old patterns. This requires a high level of training and a high vision of the common goal. Victory is not a question of collective emotion but a question of collective wisdom. When a new planetary movement takes a stand for life wholeheartedly, without reservation, then it stands on the side of higher justice and is thereby protected by high powers.

7. WATER-HEALING

Water is not merely the chemical substance H_2O, but a living organism. The new world knows the secret of water, as described most deeply by Viktor Schauberger. All the information of life from the cosmos and the earth is taken on by water and given to all beings. Healthy water, full of energy is a key to completely healing the earth. Healthy ground-water and healthy drinking water is the basis of a healthy subsistence economy, for the healing of nature and humans and for a healthy connection of new communities with the core power of life. Thanks to its strong self-cleansing powers, water can be healed relatively quickly as soon as the disruptive factors are removed and its natural forms of movement are restored. The new planetary community has started to develop new systems to heal water at several places on earth. Water can be created in even the most barren regions of the earth. If the possibilities inherent in water are used in an intelligent way, autonomous settlements can develop almost anywhere on earth. Water, energy and food are freely available for the whole of humankind!

8. THE SACRED ALLIANCE OF ALL BEINGS

The new planetary peace-community is accompanied by a host of visible and invisible co-creatures who together form the biosphere. All beings of the biosphere are in resonance with each other. Together they form a unified information system (the noosphere). This information system has been greatly disturbed by the violent interventions of the human being. Whales are losing their orientation, bees are dying out, and many dead birds have been falling from the sky recently. To heal the bisosphere again, the appropriate healing information must be introduced. All co-creatures joyfully accept information that is based on trust and respond to it enthusiastically. We know the moving pictures of the baby playing with the giant snake, of the lions who lovingly embrace their keepers, and many more. A similar co-existence with snakes, rats and wild boars has developed in Tamera. As soon as the human abandons his occult fears of the animal kingdom, the animals completely change their behaviour towards him. A cooperation starts between human and animal that previously only existed in fairytales. As soon as the first communities have developed the global information of peace, the whole animal kingdom will stand at their side. Whales and dolphins, birds, rats, frogs, ants etc. are part of an invisible information system which spreads its frequencies over the whole earth. Peace-communities of the new times will therefore do anything to rebuild the lost friendship with all creatures. This needs a radical renunciation of violence, deceit and abuse. Animals are no longer used for production. Food, cosmetics, medicines, clothes or bags for which animals had to die or suffer, will disappear from human households. The more conscientiously this happens, the stronger will be the healing power which guides the global processes from these centres.

GLOBAL SUCCESS

How can the new system of global peace spread and prevail worldwide? What gives us the optimism to so strongly believe that the global massacre will soon be over? It is an understanding of the powerful effect of new thoughts which are in resonance with the healing powers of the world. We can compare the earth's information system to a biological internet in which all the information is passed on to all participants. The system has become suffused with dense information of fear and violence, but behind this traumatic web lies a very different pattern, the pattern of healing which we call the 'Sacred Matrix'. If just a few groups on earth succeed in downloading the sacred pattern, the information of trust and healing can then be uploaded back into the global net, causing the global chain of violence to break. The healing information connects with the life forces of the sacred matrix and enters the body of the earth with great power, immediately causing genetic changes and a global field from which similar groups will arise all over the earth. A planetary process is thereby initiated, which cannot be stopped, as it is coherent with the entelechial powers of life. To illustrate this with an analogy, if we perceive the earth as a unified organism, then the healing information that is introduced has a similar effect to that of medicine introduced into the human organism. A single dose of medicine can effect a healing process in all organs and every cell! In the present case, a single dose of (complex) healing information effects a healing process in the whole organism of the earth. Human beings will no longer be physically or psychically able to torture and kill their co-creatures.

A major contribution to the shift from one age to another is provided by the power of vision. We are actually experiencing the birth of a powerful vision today: the vision of the new earth! The vision of a non-violent planet! The vision of a new planetary community! The vision of solidarity with all co-creatures! The vision of the sacred alliance! Nothing is more powerful than a vision whose time has come. If the revolutionaries of our times can develop a solid vision of peace which prevails against all resistance, then they have an unlimited power of manifestation. The power of thoughts and visions arises from the existence of the 'invisible substance'. Thoughts and visions build invisible fields of energy and information which are not limited by space. The visible world comes from invisible fields of energy and information! Just as a tree comes from the invisible substance of its genetic information. The whole of humankind is in such a process of a new becoming at present. It could be interesting in this regard to point out the famous Mayan date of December 21st 2012. We look at the scientific significance of this date, not the mythological meaning. At this point in time, different astronomical events coincide, such as a maximum of solar activity, changing the magnetic field of the earth. This change causes molecular changes in the genetic and neural structure of the human being, meaning in his consciousness and character. This process needs only relatively small triggers to lead to a genetic peace movement of planetary dimension – if these triggers are coherent with the sacred matrix. The future-trances which have been undertaken in the Peace Research Institute of Tamera, Portugal, show a shining picture. The Mayan date is not the end of humankind but a peak of global transformation and the start of a new epoch.

TAMERA AND THE GLOBAL CAMPUS

Over the past years, Tamera Peace Research Centre in Portugal has been developing a research settlement (currently 170 inhabitants) for a future without war. The thoughts of this manifesto were developed here and are being brought into reality. An international university, the so-called Global Campus, has been founded for the global dissemination of these thoughts. It has branches on several continents. The basic thoughts of the project are connected with practical developments in the fields of energy, water and food. A new model is being developed for the basics of human life, without damage to nature and our co-creatures. Energy, water and food are freely available for all of humankind, if we manage the natural resources of our earth wisely. Nobody on earth has to suffer from deprivation, starvation or cold once the tyranny is ended. May the death of so many peaceworkers not have been in vain! The call from Gaza and the earth's cry for help will no longer fade away unheard. The catastrophe in Japan has awakened millions of people. So let us come together worldwide to create a future worth living.

In the name of life.
In the name of all children.
In the name of all creatures.

THIRTY YEARS LATER
by Monika Berghoff,
Publisher Verlag Meiga, December 2011

The main features of a living cultural concept is its productive immaturity, its openness to change, self-correction and development.

The *Bauhütte* project, described in the previous chapter, led at that time to a three-year social experiment in which some important foundations for building functioning communities could be discovered and developed. The experiment turned out to be quite a success. In their enthusiasm, eight women wrote the book *Rettet den Sex* ("Saving Sex") and brought it with great courage and creativity to the public. The reaction, however, was a vigorous campaign against it – first from the media and the Church, then increasingly also from left-wing, autonomist and feminist groups, and finally from political representatives. The project and its founders were labelled as a "cult" – a defamation that in Germany means a one-way street to social isolation.

All efforts to rectify the situation had no effect. The work on the project almost came to a complete standstill. To this day, false reports are circulating on the internet, in various publications and in the records of the Church and State in Germany. The *Bauhütte* had to be dissolved. Dieter Duhm, his partner Sabine Lichtenfels and other project members left Germany to get an idea – away from the public eye – of how it could go on.

In 1991, Dieter Duhm's book Eros Unredeemed was published, and a year later his book *Politische Texte für eine gewaltfreie Erde* ("Political Texts for a Nonviolent World").

Finally in 1995 they founded, together with Charly Rainer Ehrenpreis, the Tamera Healing Biotope in Portugal. On 134 hectares of dry land with no infrastructure and only a few habitable buildings, the rebuilding of the project began – still with the firm will and aim to launch a plan for global healing.

Today, about 170 co-workers, children, teenagers, students and specialists from around the world work and live there. The founding generation has already passed most leadership tasks on to the next generation. Every year thousands of guests visit.

Various facilities and departments have arisen since the foundation: the Solar Village for the exploration and utilisation of inexhaustible energy sources, especially through a new type of solar technology requiring no photovoltaics, but also biogas plants, low-temperature Sterling engines, etc.; the Ecology Department which carries forward the idea of water and landscape healing through the implementation of water retention landscapes, permaculture, peace gardens and animal sanctuaries; the Children's Republic and Youth Place for the raising of children without fear and for a free learning; the Love School for a deep knowledge in the human realm, and for a new solidarity among women; the Political Ashram for creating a spiritual life practice in the service of global healing; the Institute for Global Peacework for a worldwide network, with a media and communications agency for the distribution of ideas, and the Global Campus as an international education initiative with base stations in various countries; the Guest and Education Centre for all those who want to know more about this cultural approach; in development are the House Akron, which will become a school of thought and a centre for arts and healing, and other departments.

The importance of this work is moving increasingly into the focus of international attention and acknowledgement. All new powers are warmly invited to collaborate and actively join the thought streams of this work.

In the name of a future worth living.

FURTHER INFORMATION

Institute for Global Peacework (IGP) • Tamera
Monte do Cerro • P-7630-303 Colos • Portugal
igp@tamera.org • www.tamera.org

LITERATURE

The titles listed below are only those closely related to the development of a new cultural idea as outlined in this book. They have been selected because of their basic impulse, their originality and their overall message, rather than due to any close agreement with the details.

Georges Bataille: *The Tears of Eros*
Ernest Callenbach: *Ecotopia*
Friedrich Nietzsche: *On the Genealogy of Morality*
Wilhelm Reich: *Character Analysis*
Wilhelm Reich: *The Function of the Orgasm*
Teilhard de Chardin: *Man and Cosmos*

only in German:
Dieter Duhm: *Der Mensch ist anders*
Dieter Duhm: *Synthese der Wissenschaft*
Hugo Kükelhaus: *Unmenschliche Architektur*
Hugo Kükelhaus: *Fassen, Fühlen, Bilden*

VERLAG MEIGA: BOOKS IN ENGLISH

Leila Dregger: *Tamera – A Model for the Future*

Dieter Duhm: *The Sacred Matrix. From the Matrix of Violence to the Matrix of Life. The Foundation for a New Civilisation*

Dieter Duhm: *Future without War. Theory of Global Healing*

Dieter Duhm: *Eros Unredeemed. The World Power of Sexuality*

Madjana Geusen (Ed.): *Man's Holy Grail is Woman. Paintings, drawings and texts by Dieter Duhm*

Sabine Lichtenfels: *Temple of Love. A Journey into the Age of Sensual Fullfillment*

Sabine Lichtenfels: *GRACE – Pilgrimage for a Future without War*

Sabine Lichtenfels: *Sources of Love and Peace. Morning Prayers*

Verlag Meiga GbR • Waldsiedlung 15 • D-14806 Belzig
www.verlag-meiga.org